Harnessing Geospatial Information Systems for Sustainable Economic and Regional Development

Arman Khan
https://orcid.org/0000-0002-8185-7350
Shaheed Benazir Bhutto University, Pakistan

Zahid Hussain
https://orcid.org/0009-0006-2182-9303
KASB Institute of Technology, Karachi, Pakistan

Abdelrehim Nader Awad
https://orcid.org/0009-0005-3649-430X
University of Bisha, Saudi Arabia

Rohit Bansal
https://orcid.org/0000-0001-7072-5005
Rockford College, Sydney, Australia

Hind Hammouch
https://orcid.org/0000-0002-5897-1649
University Sidi Mohamed Ben Abdellah, Morocco

IGI Global
Scientific Publishing
Publishing Tomorrow's Research Today

Vice President of Editorial Melissa Wagner
Director of Acquisitions Mikaela Felty
Director of Book Development Jocelynn Hessler
Production Manager Mike Brehm
Cover Design Jose Rosado

Published in the United States of America by
IGI Global Scientific Publishing
701 East Chocolate Avenue
Hershey, PA, 17033, USA
Tel: 717-533-8845 | Fax: 717-533-7115
Website: https://www.igi-global.com E-mail: cust@igi-global.com

Library of Congress Cataloging-in-Publication Data

Names: Khan, Arman, 1989- editor | Hussain, Zahid, 1996- editor | Awad, Abdelrehim Nader, 1984- editor | Bansal, Rohit, 1982- editor | Hammouch, Hind, 1991- editor

Title: Harnessing geospatial information systems for sustainable economic and regional development / Edited by Arman Khan, Zahid Hussain, Abdelrehim Nader Awad, Rohit Bansal, Hind Hammouch.

Description: Hershey, PA : IGI Global Scientific Publishing, [2026] | Includes bibliographical references and index. | Summary: "This book explores and demonstrates how geospatial technologies can be effectively utilized to support informed decision-making, promote equitable resource distribution, and foster sustainable growth across diverse regions"-- Provided by publisher.

Identifiers: LCCN 2025035295 (print) | LCCN 2025035296 (ebook) | ISBN 9798337363202 hardcover | ISBN 9798337363219 paperback | ISBN 9798337363226 ebook

Subjects: LCSH: Geospatial data--Management--Evaluation | Sustainable development | Geographic information systems--Data processing

Classification: LCC G70.217.G46 H364 2026 (print) | LCC G70.217.G46 (ebook)

LC record available at https://lccn.loc.gov/2025035295

LC ebook record available at https://lccn.loc.gov/2025035296

British Cataloguing in Publication Data
A Cataloguing in Publication record for this book is available from the British Library.

Table of Contents

Detailed Table of Contents

The pursuit of sustainable economic and regional development demands a shift from fragmented growth models toward integrated frameworks that balance environmental, social, and economic priorities. This paper proposes "Geo-Intelligence Synergies," a novel conceptual GIS framework that redefines the role of geospatial technologies in shaping sustainable futures. Unlike traditional GIS applications confined to mapping and monitoring, this framework emphasizes the synergies of spatial intelligence, predictive analytics, and policy integration. It highlights how GIS can be conceptualized as a strategic enabler of inclusive growth, equitable resource distribution, and climate-resilient regional planning. The model integrates dimensions of economic vitality, ecological stewardship, and social equity, offering a systemic approach to long-term prosperity.

The rapid digital transformation of higher education is redefining how learning is designed, delivered, and experienced. This chapter explores the intersection of smart pedagogies and intelligent campuses as key drivers of innovation in the digital age. Smart pedagogies leverage artificial intelligence, learning analytics, and immersive

Jitajitya Barh, Amity Institute of Geoinformatics and Remote Sensing, Amity University, Noida, India
K. Dhanaraj, Amity Institute of Geoinformatics and Remote Sensing, Amity University, Noida, India
Roshan Das, Department of Geography, Adamas University, Kolkata, India
Sinchan Banerjee, Department of Geography, Adamas University, Kolkata, India
Soumak Roy, Department of Geography, Adamas University, Kolkata, India
Prangya Rath, Amity Institute of Environmental Sciences, Amity University, Noida, India
Laxmi Kant Bhardwaj, Amity Institute of Environmental Toxicology, Safety, and Management, Amity University, Noida, India

Urban sprawl has emerged as a critical challenge in the context of developing nations. The KMDA region has undergone a remarkable spatial transformation over the past three decades. This study aims to assess, quantify, and analyze the spatial and temporal patterns of urban sprawl in the KMDA region from 1991 to 2021 using multi-temporal satellite data and advanced geospatial techniques. The analysis employs Landsat Thematic Mapper and OLI-TIRS imagery from 1991, 2001, 2011, and 2021 to derive LULC classifications using the supervised Maximum Likelihood method. To evaluate the spatial dispersion of urban expansion, Shannon's entropy model was employed, revealing entropy values ranging from 2.33 in 1991 to 2.95 in 2011, signifying increasing urban sprawl and spatial diffusion. The findings reveal an outward migration of the urban core from ~ 6 km in 1991 to nearly 14 km in 2021. This study will contribute to a deeper understanding of urbanization processes in rapidly growing South Asian cities.

There is increasing recognition that Geographic Information Systems (GIS) play a crucial role in shaping sustainable regional development; however, the mechanisms through which GIS adoption translates into development outcomes remain insufficiently explored in emerging economies. The present study investigates the mediating role of spatial data quality in the link between GIS adoption and sustainable regional development, and further examines how institutional capacity moderates this relationship. Data were collected from 612 respondents representing urban planning departments, municipal agencies, infrastructure authorities, and environmental management units across Karachi, Pakistan. The mean age of respondents was 34.6 years (SD = 7.83), comprising 388 males (63.4%) and 224 females (36.6%).

Sustainable tourism today is recognized not only as an environmental approach but also as a competitive strategy for attracting and retaining tourists. This study aims to examine the impact of green practices, service quality, and tourist motivation on tourist satisfaction, as well as the effect of satisfaction on the intention to revisit. The statistical population consisted of domestic and international tourists who traveled to Yazd, a cultural and sustainable destination, during the spring and summer of 2025. These tourists visited the region motivated by pursuits such as culture, history, relaxation, and responsible tourism. Data were collected using a five-point Likert scale questionnaire and analyzed through structural equation modeling using PLS3 software. The findings revealed that green practices, service quality, and tourist motivation positively influence tourist satisfaction, which in turn plays a key role in enhancing the intention to revisit.

Geospatial Information System (GIS) has come as a radical force in handling multidimensional sustainability issues within the environmental, social and economic spheres. GIS offers strong analytical to aid the planning, monitoring, and evaluation of evidence-based decisions in the countries that face the issues of climate change, increasing urbanization, resource depletion, and uneven socio-economic conditions. The role of GIS within sustainable development is considered in this chapter in an integrative manner in that it encompasses the theoretical background, empirical studies (2010-2025) and methodological aspects. It summarizes the developments in spatial analytics, remote sensing, machine learning integration, and geospatial big data management and clarifies the importance of the discussed approaches to land-use planning, disaster risk reduction, environmental conservation, renewable energy, and sustainable urban development.

Geospatial Information Systems (GIS) are increasingly serving as a fundamental facilitator of sustainable logistics by prioritizing spatial considerations in supply chain decision making. The integration of GIS with optimization techniques, remote sensing, telematics and artificial intelligence enables organizations and municipalities to enhance routing and facility placement, spatially analyze environmental and social effects, and bolster resilience against climate and infrastructure disturbances. Empirical studies indicate that GIS-based routing and dispatch systems can significantly decrease vehicle kilometers traveled, turnaround times, and carbon emissions in last-mile operations, especially when integrated with multi-objective optimization that incorporates emissions and exposure metrics. Simultaneously, GIS facilitates environmental management and energy infrastructure planning by superimposing ecological, demographic, and infrastructural layers to mitigate

Preface

Harnessing Geospatial Information Systems for Sustainable Economic and Regional Development brings together a diverse set of perspectives, research methodologies, and applied case studies that collectively demonstrate the transformative power of geospatial technologies in shaping sustainable futures. As the world confronts the complex challenges of climate change, rapid urbanization, unequal resource distribution, environmental degradation, and socio-economic disparities, the role of spatial intelligence has become more critical than ever. Geospatial Information Systems (GIS) provide more than tools for mapping, they serve as integrative platforms that unite environmental, social, and economic layers into coherent, actionable insights for policy formulation and regional planning.

This volume is the result of a collaborative effort involving scholars, practitioners, and technologists from across Asia, the Middle East, and Africa. Their contributions highlight the multifaceted applications of GIS in domains ranging from rural development, logistics management, and educational innovation to urban studies, tourism sustainability, food security, and governance. What unifies these chapters is an underlying belief: that geospatial intelligence, when effectively leveraged, can catalyze equitable development, strengthen institutional decision-making, and support resilient systems capable of withstanding emerging global threats.

The book aims to provide a comprehensive reference for academics, policymakers, planners, and professionals seeking to understand how spatial technologies can be integrated into sustainable development frameworks. Drawing on interdisciplinary research, empirical analyses, and practical case examples, the chapters collectively illuminate how GIS can guide both macro-level regional strategies and micro-level interventions that shape communities' daily experiences. By situating geospatial methods within the broader discourse on sustainability, the book underscores not only what GIS can do today, but also what it must do in the decades ahead to ensure inclusive and regenerative growth.

We express our gratitude to the contributing authors for their scholarly rigor, to our respective institutions for their support, and to the global community of GIS practitioners whose innovations continue to expand the boundaries of what is possible. It is our hope that this volume will inspire new research, foster cross-disciplinary collaborations, and empower decision-makers to harness geospatial technologies in building a more sustainable, just, and interconnected world.

CHAPTER OVERVIEW

1. Geo-Intelligence Synergies Framework for Sustainable Economic and Regional Futures

This chapter introduces the concept of "Geo-Intelligence Synergies," a forward-looking framework that expands the traditional role of GIS beyond mapping and spatial monitoring. The authors articulate an integrative model merging predictive analytics, policy design, and spatial intelligence to support climate-resilient planning and socially inclusive regional development. Emphasizing the interplay between economic vitality, ecological stewardship, and social equity, the chapter offers a theoretical foundation for reimagining GIS as a strategic enabler of long-term sustainable futures.

2. Smart Pedagogies and Intelligent Campuses: Reimagining Higher Education in the Digital Age

This chapter examines the evolution of higher education in the context of digital transformation. The authors explore how smart pedagogies—powered by AI, immersive technologies, and learning analytics—interact with intelligent campus infrastructures built on IoT, cloud systems, and smart resource management. Together, these elements form dynamic educational ecosystems that enhance equity, collaboration, sustainability, and student engagement. The chapter also discusses ethical concerns, institutional readiness, and policy considerations, laying pathways for higher education's transition toward Education 5.0.

3. Integrating Geospatial Information Systems for Sustainable Rural Development

Focusing on rural regions often underserved in terms of infrastructure, data, and technical support this chapter highlights the role of GIS in addressing long-standing developmental gaps. By integrating socio-economic and environmental data, GIS becomes a crucial platform for evidence-based rural planning. The authors elaborate on its applications in resource distribution, communication networks, agricultural management, and environmental conservation, presenting GIS as a strategic tool for fostering balanced rural development and mitigating rural–urban disparities.

4. Assessing Kolkata's Urban Sprawl: Quantifying Agglomeration Expansion Using Geospatial Technologies

This empirical study investigates three decades of urban expansion in the Kolkata Metropolitan Development Authority (KMDA) region. Using multi-temporal Landsat imagery and supervised classification techniques, the authors quantify spatial changes, analyze land-use patterns, and measure urban dispersion through Shannon's entropy. The findings reveal significant outward migration of the urban core, offering critical insights into sustainable urban planning, infrastructure development, and the challenges of managing urban sprawl in fast-growing South Asian cities.

5. Spatial Data Quality, Institutional Capacity, and GIS Adoption in Sustainable Regional Development

This chapter analyzes how GIS adoption translates into sustainable development outcomes within the context of emerging economies. Based on a survey of institutions across Karachi, the authors investigate the mediating role of spatial data quality and the moderating influence of institutional capacity. Their findings underscore that GIS adoption alone is insufficient its effectiveness hinges on data accuracy, operational readiness, administrative coordination, and organizational capability. The chapter provides policy implications for strengthening institutional frameworks in urban governance.

6. The Impact of Green Practices, Service Quality, and Tourist Motivation on Satisfaction and Revisit Intention

Centering on the tourism sector, this chapter explores how sustainability practices influence visitor experiences. Using data from tourists visiting Yazd, Iran, the study examines the role of green practices, service quality, and motivation in shaping satisfaction and revisit intention. The results highlight that tourist satisfaction functions as a critical mediator linking sustainable practices with behavioral outcomes. The chapter offers insights for tourism planners seeking to balance environmental responsibility with economic vitality.

7. Leveraging Geospatial Information Systems for Sustainable Development

This chapter provides an integrative review of GIS applications in sustainability initiatives across the environmental, social, and economic spheres. Spanning empirical studies from 2010 to 2025, it synthesizes developments in remote sensing, machine learning, spatial analytics, and geospatial big data management. The authors detail GIS applications in land-use planning, disaster risk reduction, climate adaptation, renewable energy deployment, and urban sustainability, offering a broad yet coherent synthesis of how GIS supports sustainable development agendas worldwide.

8. GIS-Enabled Logistics for Sustainable Supply Chain Management

Exploring the intersection of geospatial technologies and logistics, this chapter analyzes how GIS enhances routing optimization, facility location planning, environmental impact assessment, and supply chain resilience. Through integration with telematics, remote sensing, and AI-driven optimization, GIS enables significant reductions in emissions, travel distances, and operational inefficiencies. The chapter demonstrates how spatial intelligence strengthens supply chain sustainability while preparing logistics systems for climate-induced disruptions.

9. Harnessing GIS for Sustainable Economic and Regional Development: Integrating Spatial Intelligence into Policy, Planning, and Practice

This chapter broadens the conversation on GIS-driven regional planning by discussing how spatial intelligence can be embedded directly into governance processes. Through case examples and best practices, it illustrates the use of GIS in resource allocation, environmental monitoring, and participatory planning. The authors emphasize GIS's role in reducing regional inequalities, supporting resilient infrastructure development, and promoting spatial justice. The chapter advocates for systematic integration of geospatial tools into planning institutions to achieve balanced and inclusive growth.

10. Geospatial Applications in Right to Food and Sustainable Food Security: Global Policy Perspectives

This chapter focuses on global hunger and food insecurity, drawing on international policy frameworks such as the National Food Security Act (2013). Using

GIS-based analysis of food distribution systems, the authors assess the effectiveness of policy implementation across various states. By integrating spatial data on hunger patterns, nutrition levels, and supply chain gaps, the chapter demonstrates how geospatial tools can strengthen food security governance and crisis prevention. It also identifies policy loopholes and areas for reform, emphasizing the urgency of building geospatially informed food systems.

Chapter 1
Geo-Intelligence Synergies Framework for Sustainable Economic and Regional Futures

Yamijala Suryanarayana Murthy
https://orcid.org/0000-0002-9561-5395
KL Business School, KLEF University, India

Rohit Bansal
https://orcid.org/0000-0001-7072-5005
Rockford College, Sydney, Australia

Nishita Purthi
https://orcid.org/0000-0002-6094-0972
Asian School of Business, India

Zahid Hussain
https://orcid.org/0009-0006-2182-9303
Khadim Ali Shah Institute of Technology, Pakistan

K. Pardha Sai
https://orcid.org/0009-0001-6692-3022
Aditya University, India

ABSTRACT

The pursuit of sustainable economic and regional development demands a shift from fragmented growth models toward integrated frameworks that balance environmental, social, and economic priorities. This paper proposes "Geo-Intelligence

DOI: 10.4018/979-8-3373-6320-2.ch001

Synergies," a novel conceptual GIS framework that redefines the role of geospatial technologies in shaping sustainable futures. Unlike traditional GIS applications confined to mapping and monitoring, this framework emphasizes the synergies of spatial intelligence, predictive analytics, and policy integration. It highlights how GIS can be conceptualized as a strategic enabler of inclusive growth, equitable resource distribution, and climate-resilient regional planning. The model integrates dimensions of economic vitality, ecological stewardship, and social equity, offering a systemic approach to long-term prosperity.

INTRODUCTION

The shifting paradigm of regional development necessitates an immediate focus on spatial justice, participatory planning, and technology-oriented governance. With increasing disparities in economic, infrastructural, and service accessibility between urban and rural areas, Geospatial Information Systems (GIS) have emerged as an effective means to visualize these disparities and develop actionable solutions (Saini & Bansal, 2025). Beyond simple spatial representation, GIS technologies enable the identification of patterns of inequality, opportunity, and resilience, which are crucial for data-driven and inclusive policy action (Bansal et al., 2023).

However, conventional planning approaches often fail to capture micro-level inequities and neglect community engagement, leading to limited public ownership and technocratic decision-making (Murthy et al., 2025). In contrast, participatory governance introduces a shift toward collaborative, citizen-inclusive planning frameworks that emphasize transparency and responsiveness (Arnstein, 1969). When used alongside GIS, participatory methods enable what scholars call *geospatial governance*—the confluence of community participation, digital mapping, and institutional accountability (Yamijala et al., 2025).

Under this approach, citizens not only contribute to but also interpret and participate in the co-creation of spatial data, making the process of governance more democratic, equitable, and adaptive (Murthy et al., 2025). Such digital participation models ensure that planning outcomes reflect lived realities rather than top-down projections, thereby enhancing both trust and efficiency in policy systems (Bansal et al., 2024).

This paper explores the synergies of Geo-Intelligence—an integrated conceptual GIS framework that combines spatial intelligence, predictive analytics, and policy integration to achieve sustainable regional development. Through case studies from Telangana (India), Jakarta (Indonesia), and Nairobi (Kenya), it examines how geospatially empowered participatory governance can address spatial inequality, strengthen institutional responsiveness, and promote equitable development outcomes (Das et al., 2024; Rahman et al., 2023).

2. CONCEPTUAL FRAMEWORK

2.1 Spatial Inequality and Regional Disparities

Spatial inequality refers to the uneven distribution of resources, services, and opportunities across geographic regions, visible in disparities in healthcare, education, infrastructure, and employment. Traditional planning methods often rely on macro-level indicators that obscure localized gaps, thereby underestimating the real impact of spatial injustice (Saini & Bansal, 2025). In developing economies, public investment continues to favor urban centers, leaving peri-urban and rural regions underdeveloped (Rahman et al., 2023). This imbalance fuels migration pressures, environmental degradation, and weak institutional cohesion. Addressing such inequities requires granular, real-time data that traditional top-down assessments fail to deliver. GIS provides a powerful alternative by mapping inequality clusters and enabling data-driven, responsive policy interventions (Bansal et al., 2024; Murthy et al., 2025).

2.2 Participatory Governance: From Tokenism to Co-creation

Participatory governance represents a fundamental shift from centralized, technocratic planning toward inclusive models that empower citizens to shape policy outcomes. Drawing on Arnstein's (1969) "Ladder of Participation," most governance systems remain at intermediate rungs—consultation and placation—without actual power-sharing. Genuine participation extends beyond feedback loops, institutionalizing the right of citizens to co-create data and policy decisions (Bansal et al., 2023).

Engaging communities in data collection, spatial planning, and monitoring fosters transparency, strengthens institutional trust, and aligns policy with local priorities (Yamijala et al., 2025). Digital tools supported by artificial intelligence further democratize governance by enhancing the inclusivity and adaptiveness of civic engagement processes (Murthy et al., 2025). The integration of emotional intelligence and technological accessibility in such frameworks ensures equitable participation across socioeconomic segments (Bansal et al., 2024).

2.3 The Convergence: Geospatial Governance as an Integrative Framework

Geospatial governance merges GIS analytics with participatory practices to advance spatial justice through democratized access to spatial data (Das et al., 2024). When communities co-produce and interpret spatial information, the resulting datasets function as both analytical tools and advocacy platforms (Rahman et al., 2023). GIS-based visualizations—such as heatmaps of neglected infrastructure or geotagged citizen grievances—transform invisible inequalities into actionable insights (Murthy et al., 2025).

Empowering marginalized groups to map their realities enhances legitimacy and institutional responsiveness, while simultaneously improving data transparency and accountability (Satar et al., 2023). Such digital collaboration exemplifies emotionally intelligent, ethically grounded governance that transcends technocratic boundaries (Bansal et al., 2023; Yamijala et al., 2025).

2.4 Conceptual Model for Inclusive Regional Development

The proposed model integrates five interrelated components—spatial data layering, participatory input, analytical intelligence, governance feedback, and institutional anchoring—to promote adaptive and equitable regional development (Murthy et al., 2025).

- Spatial Data Layering: Aggregates and visualizes multidimensional inequalities to support evidence-based policymaking (Saini & Bansal, 2025).
- Participatory Input Layer: Incorporates citizen-generated data, feedback loops, and co-designed solutions (Bansal et al., 2023).
- Analytical Intelligence: Employs GIS and AI for predictive planning and resource optimization (Yamijala et al., 2025).
- Governance Feedback: Uses real-time dashboards and grievance systems to ensure policy adaptability and accountability (Murthy et al., 2025).
- Institutional Anchoring: Establishes legal mandates and cross-sectoral collaboration frameworks that institutionalize participatory GIS (Bansal et al., 2024).

This framework conceptualizes inclusive development as an iterative, learning-based process in which institutions continuously adapt through citizen feedback and spatial analytics (Das et al., 2024).

This model views inclusive development not as an endpoint, but as an iterative process—where governance learns from spatial data and continuously adapts through citizen feedback.

2.5 Barriers and Ethical Considerations

Despite its promise, geospatial governance entails ethical and operational risks. Data privacy breaches, digital exclusion, and elite capture of digital platforms can undermine equity goals (Saini & Bansal, 2025). Marginalized groups may distrust digital monitoring or lack the literacy to participate meaningfully (Rahman et al., 2023). Moreover, unregulated open-data systems risk data manipulation or misuse (Murthy et al., 2025).

Mitigation requires strong institutional safeguards such as transparent data-governance protocols, community consent frameworks, inclusive interface design, and periodic audits (Rabby et al., 2025). Without such ethical scaffolding, even sophisticated GIS applications risk reproducing existing inequalities instead of resolving them (Das et al., 2024; Bansal et al., 2024).

3. METHODOLOGY AND RESEARCH APPROACH

3.1 Rationale for Mixed-Method Design

A mixed-method design was adopted to combine the quantitative precision of GIS-based analytics with the qualitative depth of participatory governance. While GIS enables spatial identification of inequalities and infrastructural gaps, participatory methods reveal the underlying social, emotional, and institutional dimensions that influence development outcomes (Murthy et al., 2025). Integrating quantitative spatial insights with community-driven qualitative feedback ensures a multidimensional understanding of regional disparity and governance responsiveness (Bansal et al., 2023).

This approach aligns with contemporary calls for data-human synergy, wherein technology complements human judgment rather than replacing it (Rahman et al., 2023). As Murthy et al. (2025) emphasize, incorporating emotional intelligence and behavioral feedback into digital participation systems enhances both policy relevance and institutional empathy—two elements essential for equitable governance.

3.2 Case Selection Criteria

Three diverse, data-rich regional contexts were chosen for comparative analysis: Telangana (India), Jakarta (Indonesia), and Nairobi (Kenya). These cases represent varying degrees of institutional maturity, digital infrastructure, and community engagement practices (Das et al., 2024).

- Telangana was selected for its strong integration of GIS in land management, urban service delivery, and grievance systems (Murthy et al., 2025).
- Jakarta exemplifies participatory urban governance, leveraging civic-tech solutions to manage urban density and environmental risks (Mittal et al., 2025).
- Nairobi represents grassroots-driven digital innovation in mapping informal settlements and advancing spatial justice (Satar et al., 2023).

The diversity across these regions allows for both contextualized and generalizable insights. Selection was guided by five criteria: active GIS use in planning, existence of citizen-mapping initiatives, visible regional inequalities, open spatial datasets, and institutional openness to civic engagement (Rahman et al., 2023; Bansal et al., 2024).

These regions were selected based on the following criteria:

- Active use of GIS tools in policy planning or service delivery
- Existence of participatory or citizen-led mapping initiatives
- Regional development disparities evident in public records
- Availability of open-access or government spatial datasets
- Institutional openness to civic engagement in digital platforms

The geographic, political, and socio-economic diversity of these cases allows for cross-comparison and identification of context-specific as well as generalizable insights.

3.3 Data Sources

The study triangulated both primary and secondary data to ensure methodological robustness and credibility (Das et al., 2024).

A. Primary Data Collection

- Key Informant Interviews: Conducted with municipal planners, policymakers, civic-tech entrepreneurs, and NGO representatives to capture institutional perspectives (Raja et al., 2025).
- Participatory Mapping Workshops: Community-based sessions in Telangana, East Jakarta, and Nairobi's Kibera settlement gathered citizen inputs on infrastructure, access, and local priorities (Murthy et al., 2025).
- Digital Citizen Feedback Platforms: Analysis of entries from Telangana's *Janahitha* and Jakarta's *Qlue* portals to assess participatory responsiveness (Yamijala et al., 2025).

B. Secondary Data Collection

- GIS Datasets: Included OpenStreetMap, World Bank Spatial Agent, and official urban dashboards (Saini & Bansal, 2025).
- Policy Documents: Reviewed regional master plans, e-governance frameworks, and legislative mandates (Bansal et al., 2024).
- Academic Literature: Synthesized scholarship on participatory GIS, spatial inequality, and inclusive governance (Das et al., 2024; Rahman et al., 2023).

All datasets were validated using QGIS and ArcGIS tools with cross-verification to ensure accuracy and reduce duplication (Murthy et al., 2025).

3.4 Research Tools and Techniques

The analytical process integrated both quantitative GIS tools and qualitative participatory methods, promoting triangulation and depth (Bansal et al., 2023).

A. GIS-Based Spatial Analysis

- Heat Mapping: Used to visualize infrastructure density and population clusters (Saini & Bansal, 2025).
- Buffer and Proximity Analysis: Identified service deserts, healthcare coverage, and accessibility gaps (Murthy et al., 2025).
- Change Detection: Measured temporal shifts in land use and service distribution, aiding longitudinal comparison (Das et al., 2024).

B. Participatory Engagement Tools

- Community Mapping Apps: Platforms like *Maptionnaire* and *KoboToolbox* facilitated participatory spatial data collection in Nairobi and Jakarta (Satar et al., 2023).
- Geo-Tagged Grievance Systems: Telangana's digital redressal dashboard served as a model for feedback integration (Murthy et al., 2025).
- Focus Group Discussions (FGDs): Conducted to interpret citizen perceptions regarding access, equity, and institutional responsiveness (Rabby et al., 2025).

These tools enhanced participatory validity and linked digital evidence with community narratives (Yamijala et al., 2025).

3.5 Analytical Framework

The study employed a five-stage analytical integration framework to align spatial data with qualitative insights (Rahman et al., 2023):

1. Identification of Inequality Clusters: Spatial analytics pinpointed underserved areas using GIS overlays (Saini & Bansal, 2025).
2. Overlay with Community Inputs: Citizen-generated maps were integrated into official datasets, bridging data gaps (Murthy et al., 2025).
3. Policy Response Mapping: Monitored how governments responded to mapped inequities over time (Raja et al., 2025).
4. Governance Indexing: Evaluated responsiveness, transparency, and data openness (Bansal et al., 2024).
5. Cross-Case Synthesis: Compared outcomes across the three regions to derive contextual and universal insights (Das et al., 2024).

This multi-stage framework facilitated triangulation, improving reliability and interpretive validity (Murthy et al., 2025).

3.6 Limitations

Despite methodological rigor, several constraints were recognized.

- Digital Divide: Unequal internet access and digital literacy limited rural participation (Yamijala et al., 2025).
- Data Sensitivity: Restricted access to politically sensitive datasets, particularly in informal settlements (Das et al., 2024).
- Cultural Barriers: Localized facilitation was necessary to avoid miscommunication during participatory sessions (Murthy et al., 2025).

Nevertheless, triangulation through civic organizations and institutional collaboration mitigated these limitations and ensured inclusiveness (Rahman et al., 2023; Bansal et al., 2024).

4. CASE STUDY: TELANGANA, INDIA

4.1 Case Study: Telangana, India – Geospatial Governance for Equitable and Smart Development

Telangana, India's youngest state, exemplifies the transformative potential of geospatial governance in achieving inclusive growth. Leveraging advanced GIS systems, the state integrates land management, public grievance mechanisms, and service delivery within a unified digital framework (Murthy et al., 2025). With Hyderabad as its technology hub, Telangana combines political commitment and technical infrastructure to institutionalize GIS at multiple governance levels (Raja et al., 2025).

4.1.1 Background and Regional Context

Despite Hyderabad's rapid growth, rural districts such as Mahbubnagar, Adilabad, and Khammam continue to lag in infrastructure, education, and healthcare access (Saini & Bansal, 2025). Recognizing these imbalances, the Government of Telangana initiated flagship programs to promote transparency and citizen participation through digital governance platforms (Murthy et al., 2025). The Telangana State Remote Sensing Applications Centre (TRAC) and Telangana State Technology Services (TSTS) serve as institutional anchors for GIS integration (Das et al., 2024).

4.1.2 Key GIS-Driven Initiatives

A. Dharani Portal – Unified Land Records System

The *Dharani* portal merges GIS mapping with real-time property registration and grievance tracking. It enables citizens to view digitized land parcels, access ownership histories, and submit geotagged complaints, reducing bureaucratic opacity and land-related disputes (Murthy et al., 2025).

B. Urban Local Body (ULB) GIS Mapping

All 142 municipalities are mapped under integrated GIS layers, covering roads, waste collection, lighting, and water pipelines. These datasets are accessible through centralized dashboards for equitable resource allocation (Raja et al., 2025).

C. Geo-Tagged Grievance Redressal

Platforms such as *MyGHMC* and *TS-iPASS* allow real-time complaint registration with location data, significantly improving accountability and service response (Murthy et al., 2025).

D. COVID-19 Spatial Response System

During the pandemic, Telangana's GIS-enabled dashboards visualized quarantine zones and healthcare infrastructure, allowing efficient medical resource allocation (Das et al., 2024).

4.1.3 Participatory Mechanisms

Telangana stands out for embedding citizen participation within its e-governance framework. Community workshops and ward-level planning sessions empower citizens to co-identify developmental priorities (Yamijala et al., 2025). The *Janahitha* portal ensures continuous citizen feedback and transparent monitoring (Murthy et al., 2025). Such bidirectional engagement reflects the co-creation principles emphasized by Arnstein (1969) and Bansal et al. (2023).

4.1.4 Outcomes and Impacts

The integration of GIS tools has yielded measurable improvements in:

- Policy Targeting: GIS-aided prioritization of slum redevelopment and rural water projects (Saini & Bansal, 2025).
- Budget Efficiency: Reallocation of resources toward underserved wards (Das et al., 2024).
- Public Trust: Enhanced citizen satisfaction through transparent grievance tracking (Murthy et al., 2025).
- Disaster Preparedness: Improved flood forecasting and emergency planning (Rahman et al., 2023).

These outcomes highlight Telangana's success in transitioning from reactive to anticipatory governance (Raja et al., 2025).

4.1.5 Challenges and Constraints

Persistent challenges include limited digital literacy among rural populations, outdated base maps, and departmental data silos (Murthy et al., 2025). Addressing these requires institutional reforms and community-level GIS training programs (Yamijala et al., 2025).

4.2 Case Study: Jakarta, Indonesia – Crowdsourcing Spatial Justice in a Mega-City

Jakarta's urban complexity, marked by overpopulation and frequent flooding, has made it a testing ground for participatory geospatial innovation (Ajani et al., 2024). The city's Jakarta Smart City (JSC) initiative integrates GIS, citizen feedback, and social media data to drive urban responsiveness (Das et al., 2024).

4.2.1 Urban Profile and Governance Landscape

As a decentralized city-province, Jakarta's municipalities exhibit sharp intra-urban disparities in infrastructure and environmental conditions (Satar et al., 2023). The JSC program, launched in 2014, aims to close these gaps through participatory technology and open data integration (Mittal et al., 2025).

4.2.2 Key Initiatives Integrating GIS and Participation

A. Jakarta Smart City Portal

The *JSC Portal* consolidates data from 14 agencies, displaying real-time maps on mobility, waste, and pollution. Citizens submit service complaints via Twitter (X) or WhatsApp, integrating public sentiment into spatial dashboards (Bansal et al., 2024).

B. Qlue App – Participatory Governance Platform

The *Qlue* mobile application enables users to report urban issues with geotagged images. This system reduced average service resolution time from 3 days to under 24 hours (Mittal et al., 2025). It also demonstrates how digital participation can directly influence municipal performance and trust (Murthy et al., 2025).

C. Community-Based Mapping ProjectsInitiatives such as *PetaJakarta.org* utilize citizen-sourced flood data for policy intervention. By merging social media streams with official GIS data, Jakarta pioneered a model for real-time spatial governance (Satar et al., 2023).

4.2.3 Climate and Infrastructure Challenges

Jakarta's geography—40% of which lies below sea level—creates chronic flooding risks compounded by rapid urbanization (Ajani et al., 2024). Participatory GIS tools have been vital in modeling drainage systems, predicting risk, and planning adaptive infrastructure (Das et al., 2024).

4.2.4 Outcomes and Impacts

GIS-integrated participatory governance improved:

- Service Responsiveness: Qlue-enabled feedback cut response times dramatically (Mittal et al., 2025).
- Disaster Management: Real-time maps aided flood emergency coordination (Satar et al., 2023).
- Citizen Empowerment: Increased public engagement in decision-making through social platforms (Bansal et al., 2024).
- Infrastructure Planning: Data-driven prioritization of drainage and waste projects (Raja et al., 2025).

These align with Das et al. (2024), who affirm that participatory data governance enhances urban resilience and equity.

4.2.5 Challenges and Barriers

Despite success, Jakarta faces challenges such as uneven digital access, interagency data fragmentation, and the short funding cycles of civic-tech projects (Rahman et al., 2023). Moreover, mapping informal settlements can risk stigmatization or eviction without ethical safeguards (Rabby et al., 2025). Sustaining these innovations requires institutional mandates and ethical data frameworks (Murthy et al., 2025).

4.3 Case Study: Nairobi, Kenya – Grassroots Geospatial Innovation in Informal Settlements

Nairobi represents a bottom-up model of participatory GIS where community organizations lead spatial empowerment (Satar et al., 2023). Over 60% of its population resides in informal settlements, often absent from official maps or policies (Rahman et al., 2023).

4.3.1 Spatial Inequality and Informal Urbanism

Exclusion from formal spatial planning reinforces vulnerability among marginalized populations. Recognizing that *invisibility equals vulnerability*, grassroots groups began mapping unrecognized areas to advocate for equitable development (Murthy et al., 2025).

4.3.2 Key Participatory GIS and Civic-Tech Initiatives

A. Map Kibera

Launched in 2009, *Map Kibera* is Africa's first large-scale participatory mapping project, engaging local youth to document roads, schools, and water sources (Satar et al., 2023). The initiative exemplifies "data democracy" by returning data ownership to communities (Rahman et al., 2023).

B. Ushahidi Platform

Originally developed to monitor post-election violence, *Ushahidi* evolved into a multi-purpose civic platform for reporting crises, floods, and public service failures (Murthy et al., 2025). By integrating crowdsourced SMS and social media data, it bridges communication between citizens and authorities (Raja et al., 2025).

C. Spatial Collective and Open Cities Nairobi

These organizations train youth in GIS tools and collaborate with local governments on urban planning (Bansal et al., 2024). Such partnerships institutionalize participatory mapping practices within city governance frameworks (Das et al., 2024).

4.3.3 Policy Uptake and Institutional Integration

Community-generated data has directly influenced policy—Map Kibera informed county education and sanitation plans, while Spatial Collective's data contributed to land regularization efforts (Saini & Bansal, 2025). The Nairobi County Government has since adopted geo-tagged monitoring tools inspired by these models (Murthy et al., 2025).

4.3.4 Outcomes and Impacts

Participatory GIS in Nairobi resulted in:

• Spatial Visibility: Inclusion of informal settlements in master plans (Satar et al., 2023).
• Civic Empowerment: Increased youth employment and leadership (Rahman et al., 2023).
• Policy Responsiveness: Greater attention to sanitation and WASH services (Raja et al., 2025).
• Data Ownership: Community control of mapping outputs (Murthy et al., 2025).

These outcomes align with Bansal et al. (2024) and Das et al. (2024), who argue that participatory data governance enhances both resilience and institutional legitimacy.

4.3.5 Barriers and Challenges

Barriers include limited government recognition of community-generated data, funding dependence on NGOs, and risks faced by mappers documenting sensitive areas (Rabby et al., 2025). Addressing these requires stronger legal protections, open-data policies, and risk mitigation mechanisms (Murthy et al., 2025).

5. ANALYSIS AND RESULTS

Evaluating the Impact of Geospatial Governance on Inclusive Development

This section synthesizes the findings from the three case studies—Telangana (India), Jakarta (Indonesia), and Nairobi (Kenya)—to assess the effectiveness of geospatially enabled participatory governance. The evaluation framework focuses on five performance dimensions: (1) policy targeting, (2) institutional responsiveness, (3) inclusion of marginalized communities, (4) inter-institutional coordination, and (5) sustainability and ethical practice (Raja et al., 2025; Murthy et al., 2025).

5.1 Enhancing Policy Targeting and Resource Allocation

Across all three cases, geospatial governance significantly improved the precision and fairness of resource allocation. Telangana's spatial dashboards identified underserved rural wards, enabling data-driven budget reallocation (Murthy et al., 2025). Similarly, Jakarta's participatory flood maps allowed predictive planning and anticipatory budgeting for disaster mitigation (Das et al., 2024). In Nairobi, grassroots mapping revealed infrastructure deficits that guided new sanitation and education investments (Satar et al., 2023).

These outcomes validate Rahman et al. (2023), who argue that spatial intelligence, when combined with participatory inputs, enhances the contextual accuracy of policy decisions. Moreover, Saini and Bansal (2025) emphasize that equitable growth requires micro-level spatial data integration—a hallmark of all three regional frameworks.

5.2 Institutional Responsiveness and Public Trust

Integrating citizen feedback within GIS dashboards transformed governance responsiveness across cases. In Telangana, the *Janahitha* and *MyGHMC* portals allowed citizens to submit geotagged complaints and track real-time resolutions, fostering trust in public systems (Murthy et al., 2025). Jakarta's *Qlue* application reduced average complaint resolution time to under 24 hours, a tangible reflection of administrative agility (Mittal et al., 2025). Similarly, in Nairobi, citizen-led mapping initiatives improved coordination of emergency relief and sanitation services (Satar et al., 2023).

These findings align with Raja et al. (2025), who describe this as *performance-oriented governance*—a system where responsiveness and transparency are continuously reinforced through citizen interaction. Enhanced visibility and feedback loops foster a sense of co-ownership, turning government-citizen relations from hierarchical to collaborative (Bansal et al., 2024).

5.3 Spatial Inclusion of Marginalized Communities

A defining contribution of geospatial governance lies in its ability to restore visibility to previously unrecognized populations. In Telangana, peri-urban communities were included in water and sanitation planning through spatial grievance data (Murthy et al., 2025). Jakarta's flood-risk modeling integrated informal *kampung* areas, traditionally absent from official maps (Ajani et al., 2024). Nairobi's *Map Kibera* project democratized spatial visibility, transforming informal settlements into actionable zones within policy frameworks (Satar et al., 2023).

These results resonate with Das et al. (2024), who view spatial inclusion as both a developmental necessity and a moral imperative. Furthermore, Rahman et al. (2023) highlight that integrating community-generated data reshapes institutional biases, allowing equitable redistribution of infrastructure and services.

5.4 Cross-Institutional Coordination and Knowledge Sharing

The cases illustrate varying degrees of institutional integration. Telangana's *Integrated Urban Services Dashboard* exemplifies a mature governance ecosystem in which multiple departments operate on shared GIS layers (Murthy et al., 2025). Jakarta's *Smart City Portal* demonstrates effective data convergence but still struggles with departmental silos (Mittal et al., 2025). Nairobi's cross-sector collaborations, particularly through *Spatial Collective* and *Open Cities Nairobi*, promote local capacity building despite limited institutional formalization (Bansal et al., 2024).

This comparative insight reinforces Saini and Bansal (2025), who argue that cross-sectoral coordination depends on aligned data standards and shared accountability structures. Where such standards exist, as in Telangana, efficiency and policy coherence rise substantially (Raja et al., 2025).

5.5 Sustainability and Digital Ethics

While the operational success of geospatial governance is evident, long-term sustainability depends on three interrelated dimensions—digital equity, ethical safeguards, and funding stability (Rabby et al., 2025).

A. Digital Divide: In all three regions, unequal digital literacy and internet access restrict citizen participation, particularly in rural or low-income areas (Yamijala et al., 2025).
B. Data Privacy and Consent: Mapping sensitive data, such as informal housing or land ownership, can expose vulnerable groups to risks of eviction or surveillance without ethical oversight (Murthy et al., 2025).
C. Funding Models: Many participatory platforms rely on donor financing, leading to sustainability concerns once external support ends (Das et al., 2024). Telangana stands out as an exception, having institutionalized its GIS systems within state budgets (Raja et al., 2025).

Murthy et al. (2025) underscore that ethical digital design—anchored in transparency, inclusiveness, and emotional intelligence—is critical for sustaining trust in governance systems.

5.6 Comparative Synthesis

The comparative matrix below illustrates how the three cases differ in institutional maturity, participation depth, and data openness while sharing a common trajectory toward inclusive governance (Das et al., 2024; Satar et al., 2023; Rahman et al., 2023; Murthy et al., 2025).

Table 1.

Dimension	Telangana (India)	Jakarta (Indonesia)	Nairobi (Kenya)
Policy Integration	Strong: state-level GIS mandates and dashboards (Murthy et al., 2025)	Moderate: project-based coordination via Smart City Portal (Mittal et al., 2025)	Moderate: NGO–government collaboration via community mapping (Satar et al., 2023)
Citizen Participation	Moderate in rural; high in urban clusters (Yamijala et al., 2025)	High through civic apps (Qlue, Twitter) (Ajani et al., 2024)	High via grassroots mapping (Map Kibera) (Rahman et al., 2023)
Data Openness	Partial: accessible dashboards with limited API sharing (Saini & Bansal, 2025)	High: open data and crowdsourced flood maps (Das et al., 2024)	Very high: OpenStreetMap and Ushahidi platforms (Bansal et al., 2024)
Sustainability	Strong: state-funded and institutionalized (Raja et al., 2025)	Medium: dependent on hybrid public-private funding (Mittal et al., 2025)	Weak: NGO-funded and donor-driven (Satar et al., 2023)
Equity Outcomes	Strong: targeted inclusion in municipal budgeting (Murthy et al., 2025)	High: spatial inclusion in disaster and health response (Ajani et al., 2024)	High: recognition of informal settlements in policy (Rahman et al., 2023)

Overall, the analysis confirms that when spatial intelligence is aligned with participatory governance, it results in adaptive, transparent, and ethically grounded policymaking. The findings substantiate the view of Bansal et al. (2024) that inclusive digital ecosystems are fundamental to long-term regional resilience.

6. DISCUSSION

Rethinking Inclusive Governance Through Geospatial Intelligence

The convergence of geospatial intelligence and participatory governance marks a paradigm shift in development policy and practice. Beyond being a technological integration, it redefines the relationship between institutions and citizens—transforming governance from reactive administration to adaptive co-creation (Murthy et al., 2025). This section explores the conceptual and theoretical implications of geospatial governance across six critical dimensions: spatial justice, participatory inclusion, institutional transformation, trust, ethics, and scalability.

6.1 Spatial Justice as a Policy Imperative

Traditional regional development models assumed that equitable growth would result from centralized planning and trickle-down economics. However, the case studies show that spatial disparities persist without intentional interventions that make inequities visible and actionable (Das et al., 2024). GIS-based visualization democratizes information by revealing invisible biases in budget allocation and infrastructure distribution (Saini & Bansal, 2025).

As Rahman et al. (2023) argue, when inequities are made spatially explicit, they create moral and administrative pressure to address them. Geospatial tools thus operationalize *spatial justice*—a rights-based approach to planning that redistributes opportunity based on real-world geographies (Bansal et al., 2024). Consequently, GIS functions not merely as a mapping device but as an instrument of *data-driven equality* (Murthy et al., 2025).

6.2 Participatory Governance: From Inclusion to Co-Governance

Conventional participation frameworks often stop at consultation or feedback, failing to transfer real decision-making power to citizens (Arnstein, 1969). By contrast, participatory GIS (PGIS) converts citizens from passive respondents to co-producers of spatial data and policy solutions (Yamijala et al., 2025).

Examples from Nairobi's *Map Kibera* and Jakarta's *Qlue* highlight that community-generated maps and feedback systems not only enhance service delivery but also shift governance toward *co-production*—where citizens directly shape priorities and outcomes (Satar et al., 2023; Ajani et al., 2024). As Murthy et al. (2025) emphasize, participatory systems must be designed for *relational engagement*—enabling sustained, empathetic interaction rather than one-time transactions.

This aligns with Bansal et al. (2023), who note that emotional intelligence in digital systems enhances public participation by humanizing technology-mediated governance.

6.3 Institutional Transformation and Multi-Level Governance

The success of geospatial governance depends on institutional absorption and inter-agency coordination. Telangana demonstrates the transformative potential of embedding GIS into the core of state planning and budgeting systems (Murthy et al., 2025). In contrast, Jakarta and Nairobi highlight innovation from the periphery, where civic organizations and digital volunteers initiate reforms (Das et al., 2024; Satar et al., 2023).

Raja et al. (2025) identify *institutional alignment*—the coherence between local, regional, and national data systems—as the determining factor for scale and sustainability. When GIS layers are interoperable and legally mandated, as in Telangana, governance becomes anticipatory rather than reactive (Saini & Bansal, 2025). However, in contexts without formal mandates, innovations remain localized, risking fragmentation and duplication of effort.

Bansal et al. (2024) further argue that digital governance models must evolve from "pilot-based innovation" to "institutionalized transformation," ensuring continuity and accountability beyond project cycles.

6.4 Technology and Trust: A Two-Way Street

Digital tools do not automatically produce trust; they must be designed to demonstrate fairness, accessibility, and responsiveness (Murthy et al., 2025). Telangana's grievance dashboards built citizen confidence through transparency, while Jakarta's Qlue app fostered civic legitimacy via rapid response mechanisms (Mittal et al., 2025). Nairobi's community mapping projects bridged the trust deficit between marginalized populations and government institutions by validating local knowledge (Satar et al., 2023).

Krishna and Murthy (2015) argue that perceived quality of civic life is central to community well-being—an idea that resonates with the findings across all three contexts. Digital participation, when implemented transparently, strengthens relational trust, but when opaque or inaccessible, it risks reinforcing existing social divides (Rabby et al., 2025).

Trust, therefore, is both the input and the output of geospatial governance—a feedback loop where inclusive design fosters engagement, and engagement reinforces legitimacy (Rahman et al., 2023).

6.5 Ethical Dimensions of Participatory GIS

The ethical discourse surrounding GIS has expanded beyond privacy to encompass data justice, consent, and representation (Das et al., 2024). Participatory mapping initiatives may inadvertently expose vulnerable communities to surveillance or eviction, particularly when spatial data about informal settlements or contested lands is made public (Rabby et al., 2025).

Murthy et al. (2025) emphasize the need for emotionally intelligent digital design that embeds consent and cultural sensitivity into governance technologies. Ethical GIS should thus be guided by four foundational principles:

1. Informed Consent: Communities must understand the use and implications of their data.
2. Transparency: Data processes must be auditable and open for scrutiny.
3. Reciprocity: Communities contributing data should benefit directly from its outcomes.
4. Protection: Legal safeguards must prevent misuse, exploitation, or discrimination.

When these safeguards are institutionalized, GIS becomes a platform for empowerment rather than exploitation (Saini & Bansal, 2025).

6.6 Toward a Scalable and Adaptive Model

Insights from the three case studies suggest a replicable model for scalable and adaptive geospatial governance, built on five pillars: accessibility, transparency, capacity-building, institutionalization, and ethics (Murthy et al., 2025).

Table 2.

Pillar	Description
Accessibility	Design mobile-friendly, low-data platforms to ensure inclusive participation (Yamijala et al., 2025).
Transparency	Enable open dashboards, traceable data trails, and public reporting to build accountability (Bansal et al., 2024).
Capacity Building	Invest in training programs for local officials, youth, and communities to enhance GIS literacy (Das et al., 2024).
Institutional Embedding	Establish legal mandates and inter-departmental GIS units for policy continuity (Raja et al., 2025).
Ethical Design	Integrate cultural sensitivity, consent models, and fairness in system architecture (Rabby et al., 2025).

This model offers a blueprint for scaling participatory GIS across the Global South, where digital inequities and governance fragmentation persist (Rahman et al., 2023). The model not only advances efficiency but also strengthens social inclusion and accountability (Satar et al., 2023).

6.7 Theoretical Contributions

This study contributes to multiple theoretical domains within governance, technology, and spatial studies:

- Public Value Theory: Geospatial governance enhances *public value* by aligning policy outcomes with citizen-defined priorities (Bansal et al., 2024).
- Smart Governance Theory: Extends beyond e-governance by incorporating feedback loops, real-time adaptation, and participatory intelligence (Das et al., 2024).
- Spatial Citizenship: Empowers citizens as *spatial actors*—agents capable of co-producing, validating, and interpreting data for collective benefit (Murthy et al., 2025).
- Ethical Governance Frameworks: Aligns with Rabby et al. (2025), advocating for responsible data stewardship that embeds fairness and empathy.

Overall, the findings reinforce that inclusive, emotionally intelligent, and ethically anchored digital governance can transform not only administrative systems but also the citizen–state relationship itself.

7. FUTURE PATHWAYS

Charting a Sustainable and Inclusive Geospatial Future

The synthesis of findings across Telangana, Jakarta, and Nairobi demonstrates that the integration of geospatial intelligence and participatory governance can substantially improve regional equity, institutional responsiveness, and community empowerment. However, long-term success depends on how effectively these models evolve to address scalability, ethical integrity, and institutionalization challenges (Murthy et al., 2025). The following subsections outline strategic pathways for building resilient, inclusive, and adaptive geospatial governance systems in the Global South.

7.1 Institutionalizing Participatory GIS within Governance Structures

To sustain the gains of geospatial governance, participatory GIS must be formally embedded in public institutions rather than treated as pilot experiments. Telangana's model demonstrates that institutional anchoring through legal mandates and state-funded GIS dashboards ensures continuity and accountability (Murthy et al., 2025).

Local and national governments should:

- Enact legislation requiring spatial data integration in planning and budgeting (Raja et al., 2025).

• Create GIS units within municipal departments to coordinate spatial analysis (Das et al., 2024).
• Establish standardized protocols for integrating community-generated data into official systems (Saini & Bansal, 2025).
• Include participatory GIS metrics in performance evaluation frameworks (Bansal et al., 2024).

Such institutionalization mitigates fragmentation and aligns with Bansal et al. (2024), who emphasize that technological innovation achieves maximum impact only when supported by stable governance frameworks.

7.2 Investing in Digital Infrastructure and Data Equity

Geospatial governance cannot thrive without robust digital ecosystems. Persistent infrastructure gaps—such as unreliable connectivity and limited access to open spatial datasets—continue to constrain participation in rural and low-income regions (Yamijala et al., 2025).

Future strategies should prioritize:

• Development of national geospatial data infrastructures governed by open-access principles (Das et al., 2024).
• Investment in broadband expansion, satellite imaging, and interoperable platforms (Raja et al., 2025).
• Adoption of decentralized data architectures that enable local access and resilience (Murthy et al., 2025).
• Implementation of legal frameworks to ensure data equity and interoperability across institutions (Rabby et al., 2025).

As Satar et al. (2023) highlight, democratizing access to data and technological tools ensures that marginalized voices are not excluded from the digital transformation of governance.

7.3 Advancing Digital and Spatial Literacy

The sustainability of participatory GIS depends on citizen capacity to engage with spatial data meaningfully. Digital literacy gaps—especially among rural, elderly, and female populations—remain a barrier to inclusion (Yamijala et al., 2025).

To close this divide, policy efforts should focus on:

- Integrating GIS and spatial literacy into school and vocational curricula (Das et al., 2024).
- Conducting community-based mapping and data visualization workshops (Murthy et al., 2025).
- Promoting collaborations between universities, NGOs, and civic organizations for localized training (Bansal et al., 2023).
- Establishing certification programs for youth to build professional capacity in spatial governance (Saini & Bansal, 2025).

Murthy et al. (2025) emphasize that democratizing access to cognitive and emotional AI tools fosters leadership and social innovation—principles equally relevant to spatial intelligence and governance transformation.

7.4 Embedding Ethics in Design and Implementation

The future of participatory GIS must be guided by ethical design to prevent data misuse, surveillance, and exclusion (Rabby et al., 2025). As digital ecosystems expand, the risks of privacy violations, elite capture, and algorithmic bias also intensify (Das et al., 2024).

Therefore, ethical safeguards should include:

- Consent frameworks that empower communities to decide how their data is collected and shared (Murthy et al., 2025).
- "Do-no-harm" protocols for mapping vulnerable or contested areas (Saini & Bansal, 2025).
- Community data ownership and benefit-sharing mechanisms (Raja et al., 2025).
- Transparency clauses in public data policies to ensure accountability (Yamijala et al., 2025).

Rabby, Murthy, and Bansal (2025) advocate for responsible digital influence rooted in transparency and inclusivity—values that are essential for building trust and legitimacy in data-driven governance.

7.5 Scaling Innovations Through South–South Collaboration

Many pioneering participatory GIS projects, such as Map Kibera in Kenya, PetaJakarta.org in Indonesia, and Telangana's GIS dashboards, originated in the Global South (Satar et al., 2023). However, these innovations remain under-diffused across regions. To accelerate learning and replication, South–South cooperation must become a central pillar of geospatial governance policy (Das et al., 2024).

Key pathways include:

- Creating regional GIS hubs to promote knowledge exchange and capacity building (Murthy et al., 2025).
- Establishing open-source repositories for low-cost civic-tech applications (Bansal et al., 2024).
- Encouraging inter-municipal policy dialogues on participatory planning (Raja et al., 2025).
- Launching joint research platforms to study urban inequality, informality, and data governance (Rahman et al., 2023).

Satar et al. (2023) highlight that South–South collaboration fosters contextual learning and helps developing economies avoid dependency on donor-funded or proprietary systems.

7.6 Researching the Long-Term Impact of Participatory GIS

While short-term benefits—like improved responsiveness and data transparency—are well-documented, longitudinal impacts remain understudied (Rahman et al., 2023). Future research must explore how participatory GIS affects institutional behavior, political agency, and socio-economic mobility over time (Bansal et al., 2024).

Key research questions include:

- Does community-generated data influence policy beyond initial implementation cycles?
- How does long-term participation affect civic empowerment and trust?
- What are the economic returns of spatial targeting versus traditional planning models?
- How can AI–GIS convergence be managed responsibly to prevent bias and ensure fairness (Murthy et al., 2025)?

As Murthy et al. (2025) argue, integrating emotion-aware and behavior-sensitive frameworks into evaluation processes will be vital for measuring the transformative potential of digital and spatial governance. This aligns with Rabby et al. (2025), who stress that ethical and inclusive feedback systems are key to sustaining technological trust.

7.7 Conclusion to Future Pathways

In conclusion, advancing geospatial governance requires balancing innovation with institutional responsibility. The future lies in designing systems that are technologically intelligent, ethically grounded, and socially empathetic (Bansal et al., 2024). Institutional collaboration, citizen capacity-building, and ethical frameworks together ensure that GIS serves as a catalyst for justice—not merely as a technical instrument of administration (Murthy et al., 2025).

The proposed pathways contribute to global objectives under SDG 11 (Sustainable Cities and Communities) and SDG 16 (Peace, Justice, and Strong Institutions), reaffirming that spatial equity and participatory governance are inseparable pillars of sustainable development (Das et al., 2024; Rahman et al., 2023).

8. Conclusion: The findings affirm that spatial inequality is not simply a technical challenge—it is a deeply political and ethical one, rooted in histories of exclusion, invisibility, and policy neglect. Traditional planning models, limited by static datasets and top-down perspectives, often fail to capture the localized realities of underserved communities. In contrast, participatory GIS uncovers blind spots, democratizes knowledge production, and empowers citizens to not just voice grievances but co-design solutions.

Telangana's institutionalized GIS dashboards and land systems exemplify how state-level innovation can realign public investments with real-time spatial needs. Jakarta demonstrates how civic technology, even when initiated outside government systems, can drive responsiveness, transparency, and community safety. Nairobi, through grassroots platforms like Map Kibera and Ushahidi, shows how even the most marginalized communities can become visible and vocal actors in shaping urban futures. Across these cases, key success factors emerged: institutional commitment, ethical design, digital literacy, legal frameworks, and a clear feedback loop between citizen input and policy action. Equally important were partnerships—between government and civil society, between technologists and communities, and across local and international knowledge systems. However, the chapter also underscores the risks and challenges: digital divides, elite capture, data misuse, and sustainability gaps. These must be proactively addressed through inclusive design, consent-based data governance, and capacity-building efforts. Looking ahead, the integration of participatory GIS into national and regional planning frameworks offers a path toward resilient, adaptive, and just development. This approach aligns with global goals—from SDG 11 (Sustainable Cities and Communities) to SDG 16 (Peace, Justice, and Strong Institutions)—by ensuring that spatial tools serve the public good and elevate the voices of those most often left out of planning processes.

REFERENCES

Ajani, O. A., Gamede, B. T., & Bansal, R. (2024). Challenges and opportunities of integrating industry 5.0 and emerging technologies in higher education for enhancing employability skills. *Industry 5.0 and Emerging Technologies: Transformation Through. Technology and Innovation*, •••, 217–230.

Bansa, R., Bansal, T., & Pruthi, N. (2024). Demystifying the Benefits and Challenges of Virtual Learning for Differently Abled Students. In *Virtual Lifelong Learning: Educating Society with Modern Communication Technologies* (pp. 70-84). Bentham Science Publishers.

Bansal, R., Martinho, C., Pruthi, N., & Aggarwal, D. (2024). From virtual observations to business insights: A bibliometric review of netnography in business research. *Heliyon*, *10*(1). PMID: 38163120

Bansal, R., Pruthi, N., Bansal, T., & Chaudhary, A. (2023). Diving into the consumer's mind: Tools, roles, and ethical concerns of neuromarketing. In *Applications of Neuromarketing in the Metaverse* (pp. 206–222). IGI Global.

Bansal, R., Saini, S., Ngah, A. H., & Durga Prasad, T. (2024). Proselytizing the potential of influencer marketing via artificial intelligence: Mapping the research trends through bibliometric analysis. *Cogent Business & Management*, *11*(1), 2372889.

Bansal, R., Shukla, G., Gupta, A., Singh, A., & Pruthi, N. (2023). Optimizing Augmented Reality and Virtual Reality for Customer Engagement. In *Promoting Consumer Engagement Through Emotional Branding and Sensory Marketing* (pp. 24–35). IGI Global.

Das, S., Rao, N. V. J., Mishra, D., & Bansal, R. (2024). A review of the pre- and post-COVID-19 effects on the tourism and entertainment industries: Innovative methods and predicted challenges. In *Utilizing Smart Technology and AI in Hybrid Tourism and Hospitality* (pp. 98–117). IGI Global.

Krishna, K. S., & Murthy, Y. S. (2015). A study on perception of quality of work life among textile manufacturing workers in East Godavari district. [IJSR]. *International Journal of Scientific Research*, *4*(10), 1487–1491.

Krishna, K. S., & Murthy, Y. S. (2015). A study on perception of quality of work life among textile manufacturing workers in east Godavari district. [IJSR]. *International Journal of Scientific Research*, *4*(10), 1487–1491.

Krishna, K. S., & Murthy, Y. S. (2015). Employee engagement related to employees' performance. *International Journal of Engineering Technology. Management and Applied Sciences, 3*(8), 153–157.

Mittal, G., Bansal, R., Fahlevi, M., & Aziz, A. L. (2025). Mapping the research patterns on fuzzy decision-making trial and evaluation laboratory technique in marketing: A bibliometric analysis. *Social Sciences & Humanities Open, 12*, 101691.

Murthy, S. Y., Chandra, B. R., & Joshi, A. (2025). Silence is not always golden: Exploring the causes and consequences of employees' silence in the IT sector. In *Digital Transformation and Sustainability of Business* (pp. 283–288). CRC Press.

Murthy, Y. S., Bansal, R., Chodisetty, R. C. M., Chakravorty, C., & Sai, K. P. (2025). Transforming minds: AI and machine learning applications in cognitive neuroscience. In *Transforming Neuropsychology and Cognitive Psychology With AI and Machine Learning* (pp. 107–128). IGI Global Scientific Publishing.

Murthy, Y. S., Rabby, F., Bansal, R., & Chakravorty, C. (2025). Building Resilient Brands: Harnessing Influencers, Evangelists, and Data-Driven Strategies. In *Data Analytics and Influencer Marketing for Cultivating Brand Evangelism and Affinity* (pp. 71-102). IGI Global Scientific Publishing.

Murthy, Y. S., Sai, K. P., & Murthy, C. R. (2025). Integrating IoB in Smart Education: Innovative Approaches to Behavioral Data Analytics and Pedagogy. In *Internet of Behavior-Based Computational Intelligence for Smart Education Systems* (pp. 237-270). IGI Global.

. Murthy, Y. S., Srinivas, D., BS, R. C., & Kumar, G. V. (2023). Impulse Buying in the Digital Age: A Dimensional Study on Influencer Marketing Impact on Millennials. *Journal of Advanced Zoology, 44.*

Murthy, Y. S. N., Chandra, R. B., Chakravorty, C., & Shukla, S. (n.d.). Smartwatches, fitness trackers, and beyond: Evaluating experiential marketing efficacy across different age groups. In *Digital transformation and sustainability of business* (pp. 790–794). CRC Press.

Rabby, F., Murthy, Y. S., & Bansal, R. (2025). Brand evangelism in the digital era: The impact of data-driven influencer marketing. *Journal of Open Innovation, 11*(2), 100552. DOI: 10.1016/j.joitmc.2025.100552

Rabby, F., Suryanarayana Murthy, Y., & Bansal, R. (2025). Brand evangelism in the digital era: The impact of data-driven influencer marketing. *Journal of Open Innovation, 11*(2), 100552. DOI: 10.1016/j.joitmc.2025.100552

Rahman, K. T., Bansal, R., & Pruthi, N. (2023). Consumer adoption of technologies: A mindset-oriented approach. In *Contemporary Studies of Risks in Emerging Technology, Part B* (pp. 243–255). Emerald Publishing Limited.

Raja, M., Swamy, T. N. V. R. L., Das, S., Bansal, R., Fahlevi, M., & Aziz, A. L. (2025). Does human resource information system influence performance management? *Cogent Business & Management*, *12*(1), 2438862.

Saini, S., & Bansal, R. (2025). Parasocial relationships and self-disclosure: Strengthening brand image through brand attitude in premium cosmetics. *Journal of Promotion Management*, *31*(2), 337–366.

Satar, M. S., Aggarwal, D., Bansal, R., & Alarifi, G. (2023). Mapping the knowledge structure and unveiling the research trends in social entrepreneurship and inclusive development: A bibliometric analysis. *Sustainability*, *15*(7), 5626.

. Yamijala, S. M., Nimbrain, N., & Bansal, R. (2025). Enhancing remote workspaces: The role of virtual reality in shaping the virtual work environment. *Optimizing virtual reality and metaverse for remote work and virtual team collaboration*, 171-190.

Yamijala, S. M. S., Chodisetty, R. C. M., & Chakravorty, C. (2025). Adapting HR practices for a transhumanist future: An empirical study on policy and implementation. In *Enhancing the Modern Workforce Through Transhumanism* (pp. 117–146). IGI Global.

Chapter 2

Smart Pedagogies and Intelligent Campuses: Reimagining Higher Education in the Digital Age

M. Ruba
Velalar College of Engineering and Technology, India

D. Sasi Preetha
Velalar College of Engineering and Technology, India

M. Ponkarthika
Kongu Engineering College, India

V. Gomathi
https://orcid.org/0000-0002-8933-4220
Velalar College of Engineering and Technology, India

E. Sudha
https://orcid.org/0000-0002-0604-4667
Velalar College of Engineering and Technology, India

C. Pratheeba
Nandha Engineering College, India

ABSTRACT

The rapid digital transformation of higher education is redefining how learning is designed, delivered, and experienced. This chapter explores the intersection of smart pedagogies and intelligent campuses as key drivers of innovation in the digital age. Smart pedagogies leverage artificial intelligence, learning analytics, and immersive technologies to enable personalized, data-driven, and adaptive learning experiences. Intelligent campuses integrate IoT, cloud computing, and smart infrastructure to enhance academic, administrative, and environmental efficiency. Together, they form interconnected ecosystems that promote collaboration, sustainability, and inclusive education. The chapter examines theoretical foundations, technological frameworks,

DOI: 10.4018/979-8-3373-6320-2.ch002

implementation challenges, and ethical considerations, offering strategic pathways for institutions to embrace this transformation responsibly and equitably in pursuit of Education 5.0.

1. INTRODUCTION

The 21st century marks a pivotal transformation in the global higher education landscape. Universities are transitioning from traditional, physical-centric institutions toward dynamic, technology-enhanced ecosystems that emphasize innovation, connectivity, and intelligence. The widespread adoption of digital technologies, the ubiquity of data-driven decision-making, and the development of intelligent infrastructures are redefining how higher education institutions (HEIs) function, teach, and engage learners. Intelligent campuses increasingly integrate IoT-driven monitoring and autonomous control systems to enhance safety, energy management, and mobility (Ruba et al., 2023). In this context, the emergence of *smart campuses*—integrating the Internet of Things (IoT), artificial intelligence (AI), big data analytics, cloud computing, and cyber-physical systems—represents a fundamental shift in the way learning environments are designed and managed (Abu Bakar, 2025). Parallelly, *smart pedagogies*, characterized by adaptive, personalized, and experiential learning paradigms, are revolutionizing academic delivery and reshaping student engagement within the digital age (Laurillard, 2013). Higher education today is at the crossroads of two transformative forces: technological intelligence and pedagogical innovation. Traditional teaching models, rooted in lectures and standardized assessments, are increasingly giving way to data-informed, student-centered approaches that leverage analytics, machine learning, and digital platforms to personalize the learning journey. Smart pedagogies emphasize flexibility, interactivity, and contextualization, ensuring that learning is not only knowledge-based but also skill-oriented and responsive to real-world challenges. Through technologies such as AI-driven tutoring systems, learning analytics dashboards, augmented reality (AR) simulations, and virtual reality (VR) classrooms, educators are able to foster deeper engagement, critical thinking, and adaptive learning among diverse learners (Chen & Huang, 2021). At the same time, the concept of the *intelligent campus* extends beyond the digitalization of classrooms. It encompasses the holistic integration of technology into every dimension of university operations, including administration, research, infrastructure, and student life. Smart campuses leverage IoT-enabled devices to monitor and optimize energy usage, enhance campus safety, and streamline facility management. AI algorithms assist in predictive maintenance, smart mobility, and even student well-being through sentiment analysis and behavioral analytics. When combined, these technological systems create data-rich environments that support evidence-based governance and improve institutional efficiency. The intelligent campus thus becomes both a *learning space* and a *learning organism*—one that evolves continuously through data insights and human interaction.

The convergence of smart pedagogies and intelligent campuses signifies the evolution of higher education toward an ecosystemic model of learning and management. Within this model, digital infrastructure, pedagogical innovation, and human intelligence interact dynamically to create smart learning ecosystems. These ecosystems support adaptive teaching methods, personalized student services, and sustainable operations. For instance, real-time analytics derived from campus data can inform curriculum design or predict learning difficulties, enabling proactive interventions that enhance student success and retention. Similarly, faculty can use digital twins—virtual replicas of classrooms or campuses—to simulate pedagogical strategies, resource allocation, and environmental efficiency (Chen & Huang, 2021). The result is a seamless integration between digital systems and human cognition, leading to more intelligent, inclusive, and sustainable education environments.

The global momentum toward digital transformation in higher education has been further accelerated by external factors such as the COVID-19 pandemic, which forced institutions to rapidly shift to online and hybrid learning models (CARNEY, 2022). This crisis revealed both the potential and the fragility of current educational infrastructures. Institutions that had already adopted smart technologies were able to transition more effectively, while those lacking digital readiness faced significant disruptions. The post-pandemic era thus underscores the importance of digital resilience and agility in higher education systems. Smart campuses and pedagogies, when strategically implemented, can provide this resilience by enabling flexible learning pathways, remote collaboration, and automated campus management systems that can adapt to crises. However, the shift toward intelligent campuses and smart learning models is not without challenges. Ethical, technical, and organizational concerns persist regarding data privacy, algorithmic transparency, faculty readiness, and the widening digital divide (CARNEY, 2022). Questions about the ownership of student data, the fairness of AI-driven assessment systems, and the inclusivity of technology-rich environments continue to provoke debate among educators and policymakers. There is also the risk that excessive automation may erode the humanistic values of education—creativity, empathy, and critical reasoning—that cannot be replicated by machines (Laurillard, 2013). Hence, while embracing digital transformation, institutions must ensure that technological integration remains ethically grounded, inclusive, and pedagogically sound. Moreover, smart campuses and pedagogies have critical implications for sustainability and global education equity. Smart infrastructure can significantly reduce energy consumption and carbon footprints, contributing to green campus initiatives. Similarly, intelligent learning systems can democratize education by extending personalized and affordable learning opportunities to underserved regions (CARNEY, 2022). Yet, achieving this vision requires strategic investments, policy support, and interdisciplinary collaboration across academia, industry, and government. It also calls for new leadership models that champion digital literacy, ethical governance, and continuous innovation. High-performance embedded architectures play a crucial role in enabling real-time decision-making within smart learning environments (Kameswaran et al., 2017).

This chapter aims to explore how the integration of smart pedagogies and intelligent campus ecosystems reimagines higher education in the digital era. It examines the theoretical foundations and technological underpinnings of these emerging paradigms while analyzing their practical implications for teaching, learning, and institutional governance. The discussion includes an evaluation of key pedagogical innovations such as AI-driven learning, learning analytics, immersive education, and data-informed decision-making. The chapter also addresses the challenges and ethical concerns associated with smart transformations, particularly those related to inclusivity, data security, and sustainability. By synthesizing global case studies and research insights, it highlights pathways for institutions to implement intelligent systems responsibly and equitably. Ultimately, the chapter argues that the future of higher education lies in the creation of smart, connected, and adaptive ecosystems—where pedagogy and infrastructure coalesce to support holistic learning experiences and lifelong development.

2. BACKGROUND AND THEORETICAL FOUNDATION

The transformation of higher education in the digital era is deeply rooted in the convergence of technological evolution, pedagogical innovation, and systemic adaptation. As universities navigate the challenges of globalization, massification, and post-pandemic recovery, the integration of intelligent technologies has become central to achieving relevance, resilience, and inclusivity in academic systems (CARNEY, 2022). The concept of *smart education* represents an evolution beyond earlier models of e-learning and blended learning, moving toward a holistic, data-driven approach that leverages digital intelligence to personalize learning and optimize institutional operations. Understanding this transformation requires tracing the historical progression of digital education, the emergence of the smart campus paradigm, and the theoretical frameworks that underpin this convergence of pedagogy and technology. The digital turn in higher education can be traced to the late 20th century, when advancements in information and communication technologies (ICT) initiated the first wave of online learning systems. Early virtual learning environments such as Moodle and Blackboard facilitated the shift from content-centered instruction to interactive, web-based learning. However, these systems were primarily administrative in nature, focusing on course delivery rather than adaptive engagement. The current era—often referred to as *Education 4.0* or *Education 5.0*—has shifted the focus from digital access to digital intelligence, emphasizing personalized learning supported by AI, IoT, and big data analytics (Abu Bakar, 2025). This transformation parallels the principles of *Industry 4.0*, where interconnected cyber-physical systems, automation, and analytics reshape production and management processes. Similarly, in academia, smart technologies enable universities to collect, interpret, and act upon data to enhance learning, research, and administration.

The concept of the *smart campus* extends this transformation beyond pedagogy to encompass the entire university ecosystem. An intelligent campus integrates digital infrastructure, networked devices, and human-centered systems to create a responsive, sustainable, and data-informed environment. IoT devices monitor classroom conditions, lighting, and energy use; AI systems manage security, maintenance, and academic planning; and cloud-based platforms connect students, faculty, and administrators through real-time data exchange (Chen & Huang, 2021). The smart campus is thus both a *technological* and a *social* innovation—one that reconfigures spatial, administrative, and pedagogical relationships within the university. The shift toward smart campuses and pedagogies is underpinned by several theoretical perspectives that illuminate how technology mediates learning and institutional transformation. The first is Connectivism, proposed by Siemens (2005), which conceptualizes learning as a process of creating and navigating networks of information, people, and digital systems. In smart learning environments, knowledge is distributed across nodes—learners, teachers, and technologies—requiring digital literacy to connect and synthesize insights. Connectivism provides a foundation for understanding how learners engage with dynamic, data-rich ecosystems in which AI and analytics guide personalized learning pathways.

A complementary lens is Sociotechnical Systems Theory, which emphasizes the interplay between technological tools and human actors within organizational structures (Trist & Bamforth, 1951). In the context of intelligent campuses, this theory underscores that technology cannot be treated as a standalone determinant of change. Rather, the success of digital transformation depends on aligning technological systems with institutional culture, governance, and human needs (Laurillard, 2013). Faculty acceptance, digital competence, and student inclusivity become as critical as technical infrastructure in achieving smart campus functionality. This sociotechnical balance is essential to prevent technology-driven inequities and to ensure that automation enhances rather than replaces human agency. Activity Theory, derived from Vygotskian psychology, further contributes to understanding how digital tools mediate teaching and learning processes. It posits that learning is an activity system in which individuals interacts with mediating artifacts—such as AI-based tutoring systems or learning analytics dashboards—to achieve specific educational outcomes (Engeström, 2014). In a smart pedagogical context, these mediating tools reshape interactions among students, teachers, and curricula. For instance, adaptive learning platforms collect behavioral and performance data, which are then used to modify instructional strategies and feedback loops. Activity Theory thus frames smart pedagogy as a dynamic, iterative system where technology continuously adapts to learner actions. AI-powered autonomous systems have demonstrated how precision monitoring and adaptive response can enhance operational efficiency in intelligent environments (Dhanushya et al., 2025). AI-enabled visual analytics and real-time inspection frameworks can be adapted from industrial applications to optimize campus safety and sustainability (Ruba et al., 2025).

The Digital Pedagogy Framework, proposed by Laurillard (2013) also plays a critical role in conceptualizing teaching in technologically mediated contexts. It views teaching as a design science—an iterative process of creating and refining learning experiences based on technological affordances and learner feedback. Within smart campuses, this framework extends beyond online learning design to encompass the entire digital learning ecosystem, including AR/VR environments, intelligent feedback systems, and mobile learning applications. Educators act as *learning designers* who combine cognitive science, analytics, and instructional technology to produce meaningful, adaptive learning experiences. Additionally, the emerging concept of Smart Learning Environments (SLEs) integrates these theories into a unified vision of technology-enhanced education. SLEs are characterized by context-aware systems that use real-time data to adjust content, pace, and support based on individual learner needs. This adaptability not only enhances learning efficiency but also promotes inclusivity by accommodating diverse learning styles and abilities. Smart learning environments are, therefore, central to realizing Education 5.0, where human and artificial intelligence coexist to promote lifelong, equitable learning. Together, these theoretical foundations illustrate that the move toward smart campuses and pedagogies is not simply a matter of technological modernization but a paradigm shift in how knowledge is created, accessed, and applied. Universities are evolving into living, intelligent systems that merge physical infrastructure, digital platforms, and human collaboration. This shift demands an understanding of both *technical* and *humanistic* dimensions of education, ensuring that digital transformation enhances, rather than diminishes, the social and ethical purpose of higher learning. As the following sections demonstrate, the synthesis of these frameworks provides a conceptual basis for designing and implementing intelligent, inclusive, and sustainable higher education ecosystems in the digital age.

3. SMART PEDAGOGIES: REDEFINING TEACHING AND LEARNING

The paradigm of higher education is undergoing an unprecedented transformation in the digital age. Traditional, lecture-centric approaches are rapidly giving way to data-driven, adaptive, and technology-enhanced pedagogical frameworks. Smart pedagogies represent a confluence of emerging technologies, learner analytics, and innovative teaching methodologies that aim to create a more personalized, engaging, and inclusive learning experience. These pedagogies leverage artificial intelligence (AI), the Internet of Things (IoT), cloud computing, and immersive media to optimize the learning process and ensure that education evolves alongside societal and technological shifts. The shift toward smart pedagogies underscores a move from teaching to learning, from content transmission to knowledge co-creation, and from static to dynamic learning environments. Through this evolution, universities are increasingly recognizing the role of intelligent systems and analytics in augmenting human teaching capacity while ensuring meaningful student engagement and equity in digital learning spaces. Conversational AI has emerged as a transformative tool for personalizing learning pathways and offering real-time academic and emotional support (Ruba et al., 2026a). Integrating AI in pedagogy also necessitates intercultural awareness and ethical sensitivity to maintain inclusivity across diverse learning contexts (Santhini et al., 2025).

3.1. Defining Smart Pedagogies and Their Core Principles

Smart pedagogies can be broadly defined as teaching and learning strategies that employ intelligent systems, digital platforms, and real-time analytics to deliver adaptive, responsive, and personalized learning experiences (Gros & García-Peñalvo, 2016). Unlike conventional instructional methods that apply a one-size-fits-all approach, smart pedagogies dynamically adjust instructional content and assessment strategies based on learners' cognitive patterns, progress, and engagement levels. The core principles of smart pedagogies include personalization, automation, connectivity, and interactivity (Zawacki-Richter et al., 2019). Personalization ensures that learners receive content suited to their learning styles and pace, while automation reduces administrative burdens for educators by using AI-based grading, feedback systems, and content recommendations. Connectivity integrates learning environments across digital and physical spaces, creating a continuum of learning that transcends classroom boundaries. Interactivity promotes deeper engagement through simulations, gamification, and social collaboration tools. These principles align with the philosophy of Education 5.0, which emphasizes learner-centered, technology-empowered, and sustainability-oriented approaches to teaching. Smart pedagogies thus encourage not only cognitive development but also creativity, collaboration, and critical thinking—skills necessary for the future workforce in a digital society.

3.2. Artificial Intelligence in Learning and Teaching

Artificial Intelligence (AI) plays a central role in enabling smart pedagogies by facilitating adaptive and predictive learning environments. AI-driven platforms such as intelligent tutoring systems (ITS), chat bots, and virtual assistants can analyze learner data in real time to provide tailored feedback and scaffolded support (Luckin et al., 2018). These systems identify patterns in learners' performance and behavior, predicting areas of difficulty and recommending personalized interventions. For example, AI-based tutoring applications like Coursera's adaptive recommendations or platforms like Smart Sparrow dynamically adjust the difficulty level and content sequencing based on student responses. Similarly, natural language processing (NLP) tools can evaluate essays and discussion posts, offering instant feedback to enhance academic writing and communication skills (Holmes et al., 2019). In addition, AI enhances faculty productivity through automated grading systems and predictive analytics that forecast student retention risks, helping educators design timely interventions. The integration of AI not only democratizes access to personalized learning but also redefines the teacher's role as a facilitator and mentor rather than merely an information transmitter.

3.3. Data Analytics and Learning Dashboards

Learning analytics form the backbone of smart pedagogies by converting large volumes of student data into actionable insights. These analytics track attendance, engagement, assessment outcomes, and behavioral trends to generate learning dashboards that visualize student progress and identify learning gaps (Ferguson, 2012). Such insights enable educators to tailor instruction to individual needs and institutional leaders to make data-informed policy decisions. In intelligent campus ecosystems, data from smart classrooms, learning management systems (LMS), and IoT-enabled devices are aggregated to form holistic learner profiles. For instance, dashboards can highlight correlations between digital engagement and academic performance, helping to personalize interventions (Ifenthaler & Yau, 2020). However, ethical concerns around data privacy, algorithmic bias, and surveillance demand that institutions implement transparent governance and consent-based data use policies (Selwyn, 2019). Through responsible and secure data practices, learning analytics have the potential to empower both students and educators, fostering a culture of continuous improvement and reflective learning.

3.4. Immersive and Experiential Learning Technologies

Virtual Reality (VR), Augmented Reality (AR), and Mixed Reality (MR) are revolutionizing experiential learning by enabling immersive simulations and virtual classrooms. These technologies provide students with authentic, hands-on learning experiences that enhance conceptual understanding and engagement (Radianti et al., 2020). For example, VR-based laboratories allow science and engineering students to conduct experiments in risk-free virtual environments, while AR overlays digital information on physical spaces to enrich fieldwork and collaborative projects. In humanities and social sciences, immersive storytelling platforms allow students to explore historical events or social dynamics through interactive experiences. The application of these tools aligns with constructivist learning theories, which emphasize active participation and knowledge construction (Dede, 2009). As smart campuses adopt these technologies, they redefine spatial and temporal boundaries of learning, promoting global collaboration and inclusivity for remote learners.

3.5. Pedagogical Shifts Toward Learner-Centered Models

Smart pedagogies signify a broader philosophical shift toward learner-centered education, emphasizing autonomy, collaboration, and critical engagement. Blended and hybrid learning models, flipped classrooms, and problem-based learning approaches empower students to take ownership of their learning journey (Means et al., 2009). These models utilize digital tools such as adaptive quizzes, discussion forums, and cloud-based collaboration suites to create participatory learning environments. Educators transition from the "sage on the stage" to the "guide on the side," facilitating inquiry-based exploration rather than dictating knowledge (Laurillard, 2013). The result is a more inclusive and equitable educational environment that accommodates diverse learning needs and fosters lifelong learning competencies essential for Industry 4.0.

3.6. Challenges and Future Directions

Despite their transformative potential, implementing smart pedagogies presents challenges including faculty resistance, lack of digital literacy, and infrastructure disparities (Redecker & Punie, 2017). Developing nations face additional barriers related to cost, connectivity, and policy frameworks. Moreover, the ethical implications of AI-driven education—such as data privacy, intellectual property, and algorithmic bias—require robust institutional governance. Future directions emphasize human–AI collaboration, ethical AI frameworks, and universal design for learning (UDL) to ensure accessibility and inclusivity. The next generation of smart pedagogies will likely integrate cognitive computing, affective AI, and block chain-based credentialing systems to create transparent, learner-centered educational ecosystems. As higher education continues to evolve, the integration of technology and pedagogy must remain guided by humanistic values, ensuring that the purpose of education—empowerment, equity, and enlightenment—remains central.

4. INTELLIGENT CAMPUSES AND DATA-DRIVEN ECOSYSTEMS

The rise of intelligent campuses marks a revolutionary convergence between digital infrastructure, data analytics, and sustainable campus management. Unlike traditional university environments that rely heavily on manual operations and siloed data systems, intelligent campuses harness smart technologies to create responsive, adaptive, and interconnected learning and administrative ecosystems. These environments are not only designed to enhance operational efficiency but also to provide seamless, personalized experiences for students, faculty, and administrators alike. The integration of the Internet of Things (IoT), Artificial Intelligence (AI), cloud computing, and big data analytics transforms physical campuses into dynamic "living laboratories." Sensors, mobile devices, and cloud platforms generate continuous streams of data that inform real-time decision-making—from classroom energy usage to predictive maintenance, and from security management to personalized academic support. As institutions transition into these intelligent ecosystems, the focus shifts toward sustainability, inclusivity, and resilience—key values that define the future of higher education in the digital era. Ensuring cyber security and data integrity within connected campus systems remains a core challenge, addressed through intelligent detection frameworks (Krithika et al., 2025). Hardware optimization strategies have advanced to support low-latency data processing in distributed learning analytics systems (Kameswaran et al., 2017).

4.1. The Concept of an Intelligent Campus

An intelligent campus can be defined as a digitally integrated, data-driven ecosystem that optimizes academic and operational functions through smart technologies. It leverages interconnected systems—IoT devices, smart sensors, cloud-based data platforms, and AI analytics—to enable proactive management of resources and user experiences. This digital fabric connects classrooms, libraries, dormitories, and administrative offices into a cohesive, intelligent network. In practice, this means that classroom environments can automatically adjust lighting and temperature based on occupancy data, transport systems can adapt routes in real time to student demand, and campus safety systems can monitor security risks autonomously. The campus thereby becomes an intelligent organism, learning from the data it collects and evolving to improve outcomes across teaching, research, and administration. The intelligent campus also fosters collaboration between disciplines—engineering, computer science, education, and sustainability—turning the university into an ecosystem that models the "smart city" of the future (Miloslavskaya & Tolstoy, 2021).

4.2. Data Analytics as the Engine of Intelligent Campuses

At the core of every intelligent campus lies a powerful data ecosystem. Data analytics converts raw sensor and system data into actionable intelligence that supports decision-making at multiple levels. For example, predictive analytics can forecast building maintenance needs or energy consumption trends, while prescriptive analytics suggests optimal strategies for resource allocation (Williamson & Eynon, 2020). In the academic context, learning analytics are integrated with campus data to form holistic learner profiles. For instance, data on attendance patterns, digital engagement, and well-being metrics can signal potential academic risks, enabling early interventions. This integration of data systems also supports evidence-based governance, allowing administrators to assess program performance, budget efficiency, and sustainability goals in real time (Ifenthaler & Yau, 2020). However, this data-driven transformation raises critical ethical and privacy considerations. Universities must develop robust frameworks for data governance, ensuring that the collection, storage, and processing of information adhere to principles of consent, transparency, and equity (Selwyn, 2019).

4.3. IoT and Smart Infrastructure

The Internet of Things forms the technological backbone of intelligent campuses. IoT devices—ranging from smart lighting systems and connected HVAC units to RFID-enabled access cards—collect continuous data to automate campus operations (Gubbi et al., 2013). Smart classrooms, equipped with occupancy sensors, interactive boards, and environmental monitors, can adjust settings based on student presence and preferences. Moreover, IoT enhances student well-being and safety through intelligent surveillance systems, emergency alert networks, and wearable health monitors. For instance, smart wristbands can monitor health vitals in real time for vulnerable students, while environmental sensors track air quality and pollution levels within laboratories or dormitories (Chen et al., 2021). IoT-driven infrastructure also plays a vital role in sustainable campus management. Real-time energy monitoring helps reduce carbon footprints, while smart water and waste management systems contribute to environmental responsibility. These initiatives align with the United

Nations Sustainable Development Goals (SDGs), particularly Goal 11 (Sustainable Cities and Communities) and Goal 4 (Quality Education).

4.4. AI Integration for Campus Optimization

Artificial intelligence amplifies the functionality of intelligent campuses by enabling automation, prediction, and personalization. AI-powered systems can optimize class scheduling, predict equipment failures, and even recommend personalized learning resources to students (Holmes et al., 2019). Chat bots and virtual assistants provide 24/7 academic and administrative support, reducing response times and improving student satisfaction. AI-driven facility management systems can analyze historical maintenance data to schedule repairs proactively, minimizing downtime and costs. Similarly, AI-based surveillance and anomaly detection systems enhance security while minimizing human error (Luckin et al., 2018). In academic services, AI assists faculty with grading, plagiarism detection, and curriculum planning by analyzing student outcomes and engagement levels. However, to fully realize the potential of AI, campuses must invest in ethical frameworks that prevent data misuse, bias, and discrimination. The integration of human oversight ensures that technology remains a tool for empowerment rather than surveillance or exclusion (Selwyn, 2019).

4.5. Sustainability and Smart Campus Governance

Sustainability is a defining feature of intelligent campuses. Through smart energy grids, renewable energy integration, and green building technologies, campuses can drastically reduce environmental impact. AI and IoT-driven systems optimize resource consumption, while data analytics supports sustainability tracking and reporting. Governance models for intelligent campuses must emphasize transparency, accountability, and inclusivity. The integration of digital technologies in governance processes fosters real-time monitoring of institutional performance and progress toward sustainability metrics. Moreover, participatory governance models allow students and faculty to engage in data-informed decision-making, creating a shared culture of innovation and responsibility. In essence, the intelligent campus is not merely a technological advancement but a holistic transformation that aligns with broader societal goals—environmental stewardship, equity, and digital ethics.

4.6. Challenges and Future Directions

While intelligent campuses promise unprecedented efficiency and innovation, they also face significant challenges. Infrastructure costs, cyber security threats, and the digital divide remain barriers to equitable implementation (Chin et al., 2022). Data security breaches and over-reliance on automated systems can compromise trust and privacy. Future intelligent campuses will need to adopt resilient digital architectures capable of adapting to emerging technologies such as edge computing, block chain, and quantum networking. The integration of 6G and AIoT (Artificial Intelligence of Things) will enable hyper-connected, self-optimizing campuses that continuously learn and evolve. Most importantly, future development must remain human-centered, balancing technological innovation with ethics, empathy, and inclusion. Intelligent campuses should serve as laboratories for sustainable futures—where data empowers people, not replaces them.

5. INTEGRATING SMART PEDAGOGIES AND INTELLIGENT CAMPUSES- A UNIFIED FRAMEWORK

The convergence of smart pedagogies and intelligent campuses represents a transformative milestone in the evolution of higher education. While smart pedagogies focus on enhancing teaching and learning processes through adaptive and personalized approaches, intelligent campuses provide the technological and infrastructural backbone that enables these pedagogies to flourish. Together, they form a unified digital-educational ecosystem where data, intelligence, and learning environments seamlessly interact to create dynamic, learner-centered experiences. This integration signifies a shift from fragmented digital adoption to systemic innovation, where every element—students, teachers, technologies, and physical spaces—works cohesively to achieve educational excellence, sustainability, and inclusivity. The unified framework brings pedagogy, infrastructure, and data governance into alignment. It links learning analytics from classrooms with smart campus management systems to facilitate evidence-based decision-making and continuous improvement. Furthermore, it envisions the university as a *living, learning organism*—a self-optimizing system that senses, analyzes, and adapts to both pedagogical and environmental inputs in real time (Miloslavskaya & Tolstoy, 2021). Digital transformation in higher education parallels organizational agility models that align technology integration with strategic governance (Ruba et al., 2026b).

5.1. Conceptual Foundation of Integration

At its core, the integration of smart pedagogies and intelligent campuses rests on the principle of connected intelligence—the seamless flow of data and insights across academic, operational, and environmental systems (Zawacki-Richter et al., 2019). Smart pedagogies generate valuable learner data through adaptive assessments,

AI tutoring systems, and engagement analytics. Meanwhile, intelligent campus infrastructure provides contextual data about resource usage, spatial dynamics, and environmental conditions. When combined, these data streams create a 360-degree view of the learner and the learning environment. For example, IoT sensors in classrooms can measure environmental factors such as lighting and temperature, which may influence student concentration, while AI-based learning platforms capture performance patterns. Together, this data can inform personalized teaching strategies or spatial redesigns to optimize learning conditions. Thus, integration transforms education from a reactive model—responding to outcomes after they occur—into a proactive system that continuously learns and adapts, ensuring that the learning ecosystem remains responsive to both student needs and institutional goals.

5.2. Framework Components and Architecture

The proposed unified framework can be visualized as an ecosystem with four interconnected layers:

1. Infrastructure Layer – This includes IoT devices, network systems, and smart classrooms that generate real-time environmental and behavioral data.
2. Data and Intelligence Layer – Responsible for data integration, analytics, and visualization using AI, big data, and cloud computing platforms (Williamson & Eynon, 2020).
3. Pedagogical Layer – Incorporates AI-driven learning management systems, adaptive learning environments, and immersive learning technologies that apply insights from analytics to personalize education (Luckin et al., 2018).
4. Governance and Ethics Layer – Establishes data privacy, ethical use, and sustainability policies, ensuring transparency and fairness in AI-based decision-making (Selwyn, 2019).

In this architecture, data interoperability and real-time feedback loops are essential. For example, a student's participation in a virtual lab (pedagogical layer) could trigger adjustments in lab lighting or cooling (infrastructure layer) while feeding performance metrics into the learning analytics dashboard (data layer). Governance structures then ensure that these processes adhere to ethical and equitable standards.

5.3. Synergistic Benefits of Integration

The fusion of smart pedagogies and intelligent campuses creates multiple layers of value across educational, operational, and societal dimensions:

(a) Enhanced Learning Personalization: The integration allows for the continuous monitoring of learning patterns, engagement levels, and contextual factors. AI systems can tailor content delivery, recommend supplementary materials, and provide feedback suited to each learner's profile (Gros & García-Peñalvo, 2016).
(b) Predictive Academic and Operational Analytics: Institutions can leverage integrated data to forecast student outcomes, detect at-risk learners, and plan interventions before academic failure occurs (Ferguson, 2012). Similarly, predictive analytics supports facility management by anticipating maintenance needs and optimizing campus resources.
(c) Sustainability and Resource Optimization: Intelligent infrastructure integrated with pedagogical systems ensures efficient use of energy, water, and materials, aligning educational innovation with the United Nations Sustainable Development Goals.
(d) Holistic Student Experience: From personalized academic guidance to responsive learning environments, students benefit from a seamless journey that extends beyond the classroom. The integration fosters inclusivity, accessibility, and well-being, aligning with the principles of Education 5.0.

5.4. Implementation Strategies and Policy Enablers

Realizing this integration requires strategic institutional planning supported by governance and policy reforms. Universities must adopt a data-driven institutional culture, where faculty, IT teams, and administrators collaborate toward shared goals. Key strategies include:

• Developing Unified Data Platforms: Integrate learning management systems with campus data hubs to enable cross-domain analytics (Ifenthaler & Yau, 2020).
• Faculty Capacity Building: Train educators in data literacy, AI ethics, and digital pedagogy to leverage analytics responsibly.
• Collaborative Governance: Involve students and stakeholders in decision-making regarding data use and digital innovation. Public–Private Partnerships: Collaborate with technology firms for infrastructure support, research innovation, and ethical compliance mechanisms.

National and institutional policies must further ensure data sovereignty, equitable access, and long-term digital sustainability to prevent the marginalization of under-resourced institutions (Redecker & Punie, 2017).

5.5. Challenges and Ethical Considerations

Despite its promise, the integration of smart pedagogies and intelligent campuses poses challenges related to privacy, equity, interoperability, and algorithmic bias. Data silos, lack of interoperability standards, and fragmented governance hinder system-wide integration (Chin et al., 2022). Moreover, excessive reliance on automation risks reducing human agency in education. Ethical concerns around surveillance and data commodification also demand careful oversight. Institutions must ensure informed consent, transparency, and accountability in all AI-enabled educational practices (Selwyn, 2019). The development of ethical AI frameworks, grounded in fairness and inclusivity, is critical to maintaining public trust and institutional legitimacy.

5.6. Toward a Vision of Connected, Sustainable Learning Ecosystems

The integration of smart pedagogies with intelligent campuses signifies more than a technological evolution—it embodies a new educational philosophy centered on connected learning ecosystems. These ecosystems are not confined by physical or disciplinary boundaries; they are fluid, data-informed, and adaptive. In the coming decade, AIoT (Artificial Intelligence of Things) and edge computing will make real-time personalization ubiquitous, while block chain credentialing systems will enhance transparency in academic achievements (Miloslavskaya & Tolstoy, 2021). The ultimate goal is to create a sustainable, human-centered digital university—one that fosters creativity, inclusivity, and lifelong learning through responsible innovation. By bridging pedagogy and infrastructure through data intelligence, universities can reimagine themselves as ecosystems of continuous learning—where technology amplifies human potential rather than replacing it.

6. CHALLENGES, ETHICAL IMPLICATIONS, AND FUTURE DIRECTIONS

The integration of smart pedagogies and intelligent campus ecosystems has redefined the higher education landscape, but it also presents multifaceted challenges and ethical complexities that institutions must navigate. As universities accelerate digital transformation, questions of privacy, equity, accessibility, and sustainability become increasingly critical. While technologies such as artificial intelligence (AI), the Internet of Things (IoT), and learning analytics offer opportunities to personalize education and optimize campus management, they also raise concerns regarding data ownership, algorithmic bias, and the human dimensions of learning (Williamson & Eynon, 2020). Understanding these challenges and addressing them proactively is essential to ensuring that smart campuses evolve as inclusive, transparent, and ethically grounded environments. Ethical adoption of AI in education requires sensitivity to cultural communication norms and equitable assessment practices (Santhini et al., 2025). As digital ecosystems expand, the risk of cyber threats within educational infrastructures must be mitigated through AI-based threat detection systems (Krithika et al., 2025).

6.1 Data Privacy, Security, and Ownership

One of the most pressing challenges in intelligent campuses is safeguarding the massive volumes of data generated by connected systems. IoT sensors, biometric attendance systems, and AI-driven learning management tools collect sensitive information about students and faculty, including behavioral patterns, academic performance, and even emotional responses (Selwyn, 2019). Ensuring compliance with global data protection regulations such as the General Data Protection Regulation (GDPR) and maintaining transparency in data usage are paramount. Moreover, ownership of educational data remains a contentious issue—whether it belongs to the institution, the service provider, or the learner. Institutions must establish robust governance frameworks that protect privacy while enabling data-driven insights (Walton & Siemens, 2005).

6.2 Algorithmic Bias and Ethical AI

AI-powered systems are increasingly being used to support admissions, grading, and adaptive learning. However, these systems can inadvertently reinforce social and cultural biases embedded within their training data (Holmes et al., 2021). For example, predictive analytics models designed to identify at-risk students may misclassify individuals based on incomplete or biased data, leading to inequitable outcomes. Ethical AI in education requires transparency, explainability, and fairness. Universities should adopt ethical AI guidelines, ensure human oversight in decision-making, and promote algorithmic literacy among educators and administrators to mitigate unintended consequences (Zawacki-Richter et al., 2019).

6.3 Digital Divide and Accessibility

While smart pedagogies and intelligent campuses promise inclusivity, they risk deepening existing inequalities if access to digital tools and infrastructure is uneven. The digital divide—rooted in socioeconomic disparities, geographic isolation, and differing digital literacy levels—remains a barrier to equitable participation in smart education (Beaunoyer et al., 2020). To bridge this gap, higher education institutions must invest in digital infrastructure, offer training programs for both students and faculty, and develop technology-assisted learning environments that accommodate diverse learning needs, including those of students with disabilities. Accessibility must remain central to the digital transformation agenda to ensure that smart campuses empower rather than exclude.

6.4 Pedagogical and Human Challenges

Beyond technical and ethical issues, the transformation to smart campuses also affects pedagogical relationships and academic culture. Overreliance on technology may reduce human interaction, eroding the social and emotional dimensions of learning. Faculty members may experience resistance or fatigue in adapting to continuous technological change. Effective digital transformation therefore requires professional development, participatory design, and continuous feedback mechanisms to align technology adoption with pedagogical values. Educators must retain their role as facilitators of critical thinking and creativity, using technology as an enabler rather than a replacement for human connection.

6.5 Future Directions for Smart and Ethical Campuses

Looking ahead, the evolution of smart campuses and pedagogies must be guided by human-centered design, sustainability, and ethical governance. The future intelligent campus should operate as a dynamic learning ecosystem that blends physical and digital spaces, prioritizes learner well-being, and supports lifelong education. Integrating AI with immersive technologies such as virtual and augmented reality could enhance experiential learning, while block chain-based credentialing systems may ensure transparency in academic achievements. However, these advancements should be accompanied by policies that promote digital ethics, sustainability, and inclusivity. Collaborative governance models involving students, educators, policymakers, and technologists will be essential to maintaining accountability and trust. Furthermore, interdisciplinary research must continue to explore how smart technologies can support social good, reduce educational inequities, and foster sustainable campus ecosystems. Ultimately, the goal of digital transformation in higher education should not be limited to technological efficiency but should aim to cultivate resilient, inclusive, and ethically responsible learning communities (Luckin, 2018).

7. RECOMMENDATIONS AND POLICY IMPLICATIONS

As higher education transitions into the era of intelligent campuses and smart pedagogies, the need for coherent strategies, policies, and institutional frameworks becomes critical. The digital transformation of learning environments requires not only technological innovation but also ethical governance, faculty empowerment, and equitable access. Effective policy implementation can ensure that digital advancements serve the broader mission of education—fostering inclusion, sustainability, and lifelong learning rather than mere efficiency or automation. This section outlines strategic recommendations and policy implications to guide universities, governments, and stakeholders in advancing smart, ethical, and human-centered digital ecosystems. Lessons from AI-based precision agriculture demonstrate how intelligent sensing and analytics can optimize decision-making in campus sustainability (Dhanushya et al., 2025). Adapting AI-powered inspection frameworks from industrial contexts to campus environments can enhance infrastructure monitoring and predictive maintenance (Ruba et al., 2025).

7.1 Strengthening Digital Governance and Ethical Frameworks

A foundational step in enabling intelligent campuses is the establishment of comprehensive digital governance structures. Institutions must create clear policies around data ownership, privacy protection, and ethical use of AI in teaching and assessment (Walton & Siemens, 2005). Ethical frameworks should align with global standards such as the General Data Protection Regulation (GDPR) and UNESCO's AI Ethics Recommendations, ensuring that learner data is collected and used responsibly (CARNEY, 2022). Furthermore, institutions should establish independent digital ethics boards that oversee algorithmic transparency, equity, and accountability in decision-making processes. Such governance models will not only protect stakeholders but also foster institutional trust and social legitimacy in digital education.

7.2 Capacity Building and Faculty Development

The success of smart pedagogies depends largely on educators' ability to effectively use digital tools to enhance teaching and learning. Continuous professional development should be prioritized through training programs on AI integration, learning analytics, and immersive technologies (Holmes et al., 2021). Universities should encourage interdisciplinary collaboration between educators, instructional designers, and technologists to co-create adaptive learning environments. Policy frameworks must incentivize innovation in pedagogy and recognize digital competencies as core professional skills. Moreover, integrating digital literacy and ethical AI use into teacher education programs can ensure long-term sustainability of technological adoption in academia (Luckin, 2018).

7.3 Promoting Equity, Accessibility, and Digital Inclusion

Smart campuses must be built upon principles of inclusivity and universal access. Policymakers and educational leaders should address the digital divide by investing in affordable connectivity, accessible learning platforms, and assistive technologies for students with disabilities (Beaunoyer et al., 2020). Open educational resources (OERs) and low-bandwidth solutions should be promoted to ensure equitable participation, particularly in developing regions. Institutions should adopt accessibility design standards such as the Web Content Accessibility Guidelines (WCAG) and integrate social inclusion metrics in their digital transformation assessments. Equity-focused policies can ensure that smart education initiatives benefit all learners, regardless of socioeconomic background or geographic location.

7.4 Encouraging Collaboration and Interoperability

A key policy direction is to enhance collaboration across institutions, industries, and governments to build interoperable digital ecosystems. Universities should establish partnerships with technology firms, research consortia, and smart city projects to co-develop scalable and secure digital infrastructures. Shared data architectures, open standards, and collaborative platforms can enable seamless information exchange and joint innovation. Policymakers should support open innovation models and fund cross-sector research to foster experimentation in emerging technologies such as AI-driven learning analytics, block chain credentialing, and digital twins for campus management. Interoperability is vital to preventing technology silos and ensuring sustainable scalability across the higher education sector.

7.5 Integrating Sustainability and Lifelong Learning Agendas

Digital transformation must align with broader sustainability and lifelong learning goals outlined in the United Nations Sustainable Development Goal 4 (Quality Education). Smart campuses should adopt energy-efficient infrastructures, green data centers, and AI-enabled resource optimization systems to minimize environmental impact (CARNEY, 2022). Policymakers should also promote micro-credentialing, modular learning, and flexible digital pathways that support continuous upskilling and reskilling in a rapidly evolving workforce. Integrating sustainability principles into campus operations and pedagogy can position higher education as a catalyst for global digital equity and ecological stewardship.

7.6 Policy Integration and Future Vision

Ultimately, digital transformation in higher education demands a multi-stakeholder, whole-system approach. Governments must integrate education, digital economy, and innovation policies to create cohesive frameworks that empower institutions to evolve into intelligent campuses. Accreditation agencies and quality assurance bodies should update their evaluation criteria to include digital pedagogy standards and ethical AI practices. The future intelligent campus should not only leverage technology for operational excellence but also nurture human creativity, critical thinking, and social responsibility. Through inclusive policies and ethical innovation, higher education can truly reimagine its purpose in the digital age—preparing learners to thrive in a smart, sustainable, and interconnected world.

8 CONCLUSION

The convergence of smart pedagogies and intelligent campus ecosystems marks a transformative moment in the evolution of global higher education, redefining how knowledge is created, delivered, and experienced in the digital age. As universities transition from traditional, lecture-centric paradigms to dynamic, data-driven, and student-centered learning environments, the boundaries between physical and digital spaces are dissolving to create integrated smart learning ecosystems (Williamson & Eynon, 2020). Smart pedagogies—rooted in personalization, interactivity, and adaptability—empower learners to become active participants in their educational journeys through AI-driven tutoring systems, immersive virtual simulations, and analytics-based feedback mechanisms that continuously optimize learning outcomes (Holmes et al., 2021). Simultaneously, intelligent campuses, enabled by the Internet of Things (IoT), big data analytics, and cloud computing, are reshaping institutional operations by embedding intelligence into infrastructure, energy management, security, and learning environments. Together, these innovations herald a new paradigm—often termed Education 5.0—where technology is not merely a tool but an integral enabler of human potential, fostering creativity, collaboration, and lifelong learning (Luckin, 2018). The future of higher education will increasingly rely on conversational AI systems that humanize digital learning experiences and democratize access (Ruba et al., 2026a). However, this digital transformation is not without challenges. Issues of data privacy, algorithmic transparency, digital inequity, and ethical AI deployment remain central to the discourse, necessitating the development of robust governance frameworks and institutional policies that prioritize human agency, inclusivity, and sustainability (Walton & Siemens, 2005).

The future intelligent campus must therefore operate within a human-centered ethical framework that balances technological efficiency with social responsibility and academic integrity. Furthermore, as educational institutions increasingly adopt AI, block chain, and extended reality technologies, it becomes essential to ensure that their deployment aligns with pedagogical goals rather than commercial imperatives, maintaining education as a public good accessible to all learners regardless of geography or socioeconomic status (CARNEY, 2022). The role of educat The interdisciplinary synergy between technology, pedagogy, and organizational design will define the next phase of intelligent educational transformation (Ruba et al., 2026b). ors will continue to evolve from transmitters of information to facilitators of knowledge ecosystems, curators of digital experiences, and mentors who guide learners through complex networks of information and technology. Professional development and digital literacy training will remain essential for empowering faculty to navigate emerging technologies while retaining the emotional, creative, and ethical dimensions of teaching.Moreover, higher education must embrace interdisciplinary collaboration—bridging computer science, cognitive psychology, design, and educational theory—to ensure that smart pedagogical models are inclusive, evidence-based, and adaptable across diverse contexts. From a policy perspective, the future of intelligent campuses demands multi-stakeholder collaboration among governments, technology providers, educators, and learners to co-develop open, interoperable, and sustainable digital infrastructures. Investment in accessible digital resources, equitable bandwidth distribution, and open educational platforms will be vital to bridging the global digital divide and fostering educational justice in both developed and developing regions (Beaunoyer et al., 2020).

Additionally, the integration of sustainability principles into smart campus design—such as green computing, energy-efficient buildings, and smart environmental monitoring—can align digital transformation with global environmental goals, reinforcing higher education's contribution to the United Nations Sustainable Development Goals (CARNEY, 2022). Looking ahead, the intelligent campus of the future will likely evolve into a fully connected cognitive ecosystem where AI-driven analytics anticipate learning needs, augmented reality facilitates immersive knowledge construction, and block chain ensures transparent credentialing. Yet, despite technological advancements, the ultimate success of smart higher education will depend not on machines but on the values that guide their use—equity, empathy, and ethics. As emerging technologies like quantum computing, digital twins, and generative AI become increasingly prevalent, universities must continuously reassess their policies, pedagogical models, and ethical standards to ensure that innovation serves the collective good rather than exacerbating inequalities or eroding human connection. In essence, the future of higher education lies not merely in digitizing traditional processes but in reimagining the very foundations of learning, scholarship, and institutional purpose. By harmonizing smart pedagogies with intelligent campus infrastructures, universities can cultivate ecosystems that are resilient, adaptive, and inclusive—ecosystems that nurture intellectual curiosity while addressing the evolving needs of society and the planet. As the digital age continues to unfold, the challenge for educators, policymakers, and learners alike is to shape an educational future that is as intelligent as it is humane, where technology amplifies human creativity, data empowers ethical decision-making, and smart campuses become living laboratories for sustainable, equitable, and lifelong learning (Zawacki-Richter et al., 2019). Thus, the journey toward truly intelligent higher education is not solely a technological pursuit but a moral, social, and intellectual endeavor—one that invites us to redefine what it means to learn, teach, and innovate in an increasingly interconnected and intelligent world.

REFERENCES

Abu Bakar, M. R. (2025). A systematic review of smart campus initiatives in transforming higher education institutions. [JIKM]. *Journal of Information and Knowledge Management*, *15*(SI2), 176–188.

Beaunoyer, E., Dupéré, S., & Guitton, M. J. (2020). COVID-19 and digital inequalities: Reciprocal impacts and mitigation strategies. *Computers in Human Behavior*, *111*, 106424. PMID: 32398890

Carney, S. (2022). Reimagining our futures together: a new social contract for education: by UNESCO, Paris, UNESCO, 2021, 186 pages, ISBN 978-92-3-100478-0.

Dede, C. (2009). Immersive interfaces for engagement and learning. *Science*, *323*(5910), 66–69. PMID: 19119219

Dhanushya, J., Ruba, M., Ponkarthika, M., Soundarya, B., Sakthi Priya, V., Dhanushri, S., & Harikrishnan, S. (2025, April). AI-Powered Multipurpose Agrobot for Precision Farming: Crop Health Management, Fertilizer Recommendation and Yield Prediction. In *2025 International Conference on Recent Advances in Electrical, Electronics, Ubiquitous Communication, and Computational Intelligence (RAEEUCCI)* (pp. 1-6). IEEE.

Engeström, Y. (2014). *Learning by expanding: An activity-theoretical approach to developmental research*. Cambridge University Press.

Ferguson, R. (2012). Learning analytics: drivers, developments and challenges. International journal of technology enhanced learning, 4(5-6), 304-317.

Gros, B., & García-Peñalvo, F. J. (2016). Future trends in the design strategies and technological affordances of e-learning. In *Learning, design, and technology* (pp. 1–23). Springer.

Gubbi, J., Buyya, R., Marusic, S., & Palaniswami, M. (2013). Internet of Things (IoT): A vision, architectural elements, and future directions. *Future Generation Computer Systems*, *29*(7), 1645–1660.

Holmes, W., Bialik, M., & Fadel, C. (2019). *Artificial intelligence in education promises and implications for teaching and learning*. Center for Curriculum Redesign.

Kameswaran, A., Nanthakumar, K., & Ruba, M. (2017). High performance precomputation based self-controlled precharge-free content-addressable memory. *Journal of Electronics and Communication Systems*, *2*, 1–9.

Krithika, S., Jayachandran, T., Raghapriya, N. R., Ruba, M., Janani, C., & Sathiyavathi, S. (2025, April). An effective denial of service texture features from malware grayscale image. In *Proceedings of the 2025 3rd International Conference on Artificial Intelligence and Machine Learning Applications (AIMLA)* (pp. 1–7). IEEE.

Laurillard, D. (2013). *Teaching as a design science: Building pedagogical patterns for learning and technology*. Routledge.

Luckin, R. (2018). Machine Learning and Human Intelligence. The future of education for the 21st century. UCL institute of education press.

Means, B., Toyama, Y., Murphy, R., Bakia, M., & Jones, K. (2009). Evaluation of evidence-based practices in online learning: A meta-analysis and review of online learning studies.

Radianti, J., Majchrzak, T. A., Fromm, J., & Wohlgenannt, I. (2020). A systematic review of immersive virtual reality applications for higher education: Design elements, lessons learned, and research agenda. *Computers & Education, 147*, 103778.

Redecker, C., & Punie, Y. (2017). *Digital competence of educators* (Punie, Y., Ed.).

Ruba, M., Ashwin Prabhu, G., Deka, J., Kavitha, D., Sathiya, S., & Pavithra, S. (2026). Empowering learners and educators: The transformative role of conversational AI in personalized education and student support. In Assayed, S. (Ed.), *Enhancing student support and learning through conversational AI* (pp. 163–190). IGI Global Scientific Publishing., DOI: 10.4018/979-8-3373-3316-8.ch007

Ruba, M., Ashwin Prabhu, G., Robinson Joel, M., Vishnu Priya, K., Dharshani, L., & Dharani, R. (2026). Integrating emerging technologies for organizational agility: An interdisciplinary approach to digital transformation. In Roache, D. (Ed.), *High performance leadership for organizational excellence: Innovation, technology, and resilience* (pp. 63–92). IGI Global Scientific Publishing., DOI: 10.4018/979-8-3373-6207-6.ch003

Ruba, M., Girinath, V. P., Ponkarthika, M., Santhini, C., Sakthi Priya, V., & Hemalatha, R. (2023, December). Design and implementation of smart internet-based real-time control of rover with artificial intelligence. In *Proceedings of the 2023 International Conference on Energy, Materials and Communication Engineering (ICEMCE)* (pp. 1–5). IEEE. https://doi.org/DOI: 10.1109/ICEMCE2023

Ruba, M., Ponkarthika, M., Parameshwari, V., Girinath, V. P., Soundarya, B., Janani, M., & Santhini, C. (2025, February). Enhancing underwater tunnel safety through smart inspection: An ROV-based system with AI-powered crack detection using Canny edge detection. In Proceedings of the 2025 3rd International Conference on Intelligent Systems, Advanced Computing and Communication (ISACC) (pp. 1067–1072). IEEE. https://doi.org/DOI: 10.1109/ISACC.2025

Santhini, C., Ruba, M., Sakthi Priya, V., & Hemalatha, R. (2025). Intercultural communication and assessment practices in the integration of AI with education. In *Generative AI Approaches to Sustainable Development in Higher Education.* IGI Global.

Selwyn, N. (2019). *Should robots replace teachers?: AI and the future of education.* John Wiley & Sons.

Siemens, G. E. O. R. G. E. (2005). Connectivism: A learning theory for the digital age. International Journal of Instructional Technology and Distance Learning. On-line] retrieved from: http://www. idtl. org/Journal/Jam _05/article01. html

Williamson, B., Eynon, R., & Potter, J. (2020). Pandemic politics, pedagogies and practices: Digital technologies and distance education during the coronavirus emergency. *Learning, Media and Technology*, *45*(2), 107–114.

Zawacki-Richter, O., Marín, V. I., Bond, M., & Gouverneur, F. (2019). Systematic review of research on artificial intelligence applications in higher education–where are the educators? *International Journal of Educational Technology in Higher Education*, *16*(1), 1–27.

Chapter 3
Integrating Geospatial Information Systems for Sustainable Rural Development

Ishu Aggarwal
Vivekananda Institute of Professional Studies-TC, Guru Gobind Singh Indraprastha University, New Delhi, India

Aashu Aggarwal
https://orcid.org/0000-0003-0812-1504
Vivekananda Institute of Professional Studies-TC, Guru Gobind Singh Indraprastha University, New Delhi, India

ABSTRACT

Despite being the crucial resource centres of the nation capable of providing crucial and strategic resources, rural areas have largely been found struggling with a lack of advanced infrastructural facilities, strong communication networks, proper information systems, and up-to-date technical assistance considered necessary for their sustainable growth and development. In light of these challenges and the continuing deteriorating state of rural areas, it becomes imperative to plan policies and strategies for sustainable rural development that would help foster balanced economic growth and environmental conservation. Of all the technological advancements, Geospatial Information Systems (GIS) have emerged as a strategic tool that combines geographic data with socio-economic and environmental information to provide a reliable and effective solution for logical and informed decision-making. This chapter takes forward this discussion to get deeper insights into GIS and its integration in sustainable rural planning and development.

DOI: 10.4018/979-8-3373-6320-2.ch003

1. INTRODUCTION

Rising population combined with increasing rural-urban migration, adverse climate changes, increasing environmental degradation, and increasing socio-economic imbalance has significantly hampered the rural areas, tarnishing the quality and efficiency of these rural areas and their activities. Despite sustainability in talked about for so many years, rural areas have remained largely neglected when addressing the economic growth of the nation. With greater advancements in technology, as the urban areas get developed, rural areas continuously face stagnation, with their processes becoming obsolete and inefficient. Regrettably, even after quite a bit of time, where sustainable planning has been mentioned, it continues to be overlooked, particularly when it comes to rural areas. The lack of analytics or information on analytics makes addressing the rising concerns in rural areas even more difficult. Given the slower growth in rural areas, innovative, data-driven decision-making, is essential to address the rising challenges, like resource depletion, climate change, population growth, etc. Given that conventional rural planning relies on outdated data, mission creep, and the use of theory that defies rational explanation is irrelevant to these pressing problems, researchers can find a way to develop area-defined solutions for sustainable rural development. Within this context, geospatial information technology can help with the administration, analysis, and storage of spatial data. The existing geospatial information system (GIS) can facilitate the kind of informed and rational decision-making that can promote sustainable and beneficial rural planning. Given these benefits, policy-makers, corporations, and communities are increasingly utilizing these systems to visualize spatial relationships, utilize resources more effectively, and facilitate sustainable development. For example, remote sensing devices such as satellites and aerial photography routinely supply data about land use and land cover (Deichmann & Wood, 2001). Further, precision farming, a nascent agricultural practice, is also based on these geospatial technologies (Mandal & Ghosh, 2000). The government of India has notified the National Geospatial Policy, 2022, as a transformative measure towards positioning India as a global leader in the geospatial sector. The policy, with its long-term vision, seeks to foster geospatial technology adoption, inclusion, and economic growth by liberalizing access to geospatial data. As businesses and policy leaders become much more involved in the planning and implementation of sustainable development (SD) projects - specifically in rural areas, to help achieve the Sustainable Development Goals (SDGs) - (Hossain et al., 2023), combining geographic information systems (GIS) with sustainable rural development, presents an innovative framework for inclusive and precise planning, monitoring, and evaluation of development projects. Notwithstanding the above significant advancements in the field of GIS, there is still some unawareness, lack of appreciation, and acceptance of the value of geospatial information and its supporting architecture especially in policy and decision-making. (Scott and Rajabifard, 2017).

1.1 Objectives of the Study

The objective of this chapter is to explore the concept of Geospatial Information Systems (GIS) in detail while carefully addressing its integration and surrounding challenges in reference to sustainable rural development. The chapter is an attempt to gain deeper insights into the applications and implications of GIS for the sustainable development of rural areas. Additionally, the chapter also details the strategies to overcome the challenges encountered in the successful integration of GIS. We believe that the discussion would significantly contribute to the literature, enabling policymakers and strategists to utilize GIS to its maximum potential and with the greatest efficiency and effectiveness. The study attempts to answer the following research questions:

RQ1. What are Geospatial Information Systems (GIS) and Sustainable Rural Development?
RQ2. What are the applications and implications of GIS for sustainable rural development?
RQ3. What are the challenges involved in the integration of GIS for sustainable rural development?

2. LITERATURE REVIEW

2.1 Geospatial Information Systems (GIS)

The idea of a Geographic Information System (GIS) had gained widespread acceptance by businesses, educational institutions, government organizations, and private citizens by the late 1970s. The first likely accurate GIS was created by Dr. Roger Tomlinson (1933–2014), who is recognized as the father of GIS. At the simplest stage, a GIS may often be considered to be the high-tech equivalent of any map. A computer-based information system for managing geographical data and applying that data to address spatial issues is called a Geospatial Information System (GIS) (Lo and Yeung, 2007), and geospatial data are those that have spatial and geographic orientations and contain information content with a location component (Meeks and Dasgupta, 2004). The initial ideas of geospatial information were established in the early 1960s to report on environmentally sustainable development and the integration of statistics and mapping data. GIS has moved from the desktop to the palm of your hand thanks to a mix of newer operating system versions, more powerful computers, and portable devices. The power of GIS is now available to people worldwide, allowing for more dynamic, real-time, and location-aware interactions with their surroundings. The development and use of geodetic and information science techniques for modeling, gathering, exchanging, organizing, exploring, analyzing, synthesizing, visualizing, and assessing data on Earth-related spatiotemporal phenomena is the focus of geospatial information science (Bill and others, 2022). Today, with rising challenges, there is no way to go back to the past since the world that we humans have built and are currently in is so different from any previous reality. Geospatial technology can help with these issues, particularly if it is developed properly, is sufficiently extensive, and is available to everyone. Many forms of geographic data could now be efficiently captured thanks to the availability of off-the-shelf technologies and basic data models. Today, geospatial technology is used in almost every aspect of human endeavor, aiding in problem-solving, decision-making, outcome prediction, and the understanding and explanation of the functioning of the social and environmental systems on Earth. Through the use of maps and data, geospatial technology enables its users to "see what others can't," revealing information that would not otherwise be apparent, valued, or comprehended (Dangermond and Goodchild, 2020). Finding cost-effective information that is thematically, geographically, and chronologically relevant to support policy analysis and decision-making is one of the biggest issues in development planning (Schultink, 2000). Further, the safeguarding and archiving of data in an immediately accessible digital-formatting platform not only makes it possible to print hard-copy maps much more quickly and economically, but it also makes complicated analysis and modeling possible that was before impossible (Abdalla, 2016).

2.2 Sustainable Rural Development

Rural areas are often found struggling with access to essential facilities, such as education and healthcare, thereby hindering their development. Lack of empirical information, concerns about scientific knowledge, and disputes over the management of knowledge are some of the key issues in studying rural development within the context of sustainability discourse (Bruckmeier and Tovey, 2008). Rural development has become essential for achieving the SDGs (Ashley and Maxwell, 2001). It is a concept that focuses on the development of rural areas while addressing the issues of scarcity of resources. As rural areas show a new spatial-economic dynamics, it becomes imperative to use economic-ecological sustainability as a core principle for new rural development (Akgün et al., 2015). Sustainable rural development centers on balanced growth, covering economic viability, social inclusion, and environmental care. The objective is to create synergies among various development sectors, such as energy, health, education, water, food, etc (Hossain, 2017). Today, rural regions are benefiting from the use of modern technology and economic incentives to enhance social benefits, markets, employment, income generation, natural resource management, and environmental conservation (Shahid et al., 2024). Advances in digital technology and innovative solutions now enable rural businesses to directly source and sell goods within their communities, strengthening the local economy and addressing issues of inequality and poverty. Rural areas are especially interested now in improving current methods for assessing and managing rural settlement systems.

2.3 Geospatial Information Systems and Sustainable Rural Development

Information and Communication technologies (ICT) have emerged as a contemporary tool to facilitate rural planning and management. ICT contributes to successful and sustainable rural development strategies through improved access to information, enhanced communication, and effective decision-making. The achievement of sustainable development goals (SDGs) to a large extent depends on the successful implementation of remote sensing and geospatial (geographic) information systems (Dwivedi et al., 2024). GIS encompasses the collection, integration, visualization, and exploration of contextual and geographic data and spatial information (Blasch et al., 2011). Rural resilience, circular economy, and the right education become the strategic guides to develop rural economies and reduce socio-economic inequalities and environmental justice (Mihai and Iatu, 2020). As explained by Kurowska (2020), sustainable rural development is a conscious transformation process that requires aligning human needs with environmental needs. In such a scenario, GIS tools help in rapid and comprehensive analysis of the environment and spatial phenomena by helping to disseminate reliable information. These systems and tools are essential to development forecasting and sustainable rural development planning. Recent applications of GIS include automated farming systems, precision farming, climate change detection, and real-time monitoring of agricultural production (Karunathilake et al., 2023). Geospatial technology is influential in changing the agricultural sector by using remotely sensed images to implement good agricultural practices (Pandey and Pandey, 2023).

3. RESEARCH METHODOLOGY

The present chapter is an exploratory cum descriptive study that attempts to describe the applications of GIS. The study builds on the secondary data collected from various sources, such as research papers, journals, and global databases. The diverse sources of information were considered to ensure a comprehensive discussion of the topic.

4. DISCUSSION

4.1 Geospatial Information Systems (GIS) in Rural Planning

Geospatial Information Systems are often referred to as GIS. They have become essential for planning in rural and remote areas. These systems let planners and decision-makers look at how things connect across various regions and spaces. Researchers find them helpful too. They spot environmental changes pretty easily. Then they come up with strategies for sustainable development. In rural areas, folks apply GIS in all sorts of ways. That includes land use planning and agricultural management. It covers infrastructure development along with environmental monitoring (FAO, 2017). The tool helps pinpoint the best spots for farming. It maps out resources in detail. Plus, it locates ideal sites for services like education and healthcare. Transportation fits in there as well (Hargreaves & Watmough, 2021). When it comes to precision agriculture, GIS maps soil types carefully. It tracks water sources and crop yields too. All this boosts productivity. At the same time, it supports environmental sustainability in a real way. GIS also makes community planning more inclusive. Interactive maps encourage people to share knowledge. They join in on decision-making processes (Fallah & Karami, 2023). The system plays a part in rural tourism as well. It maps natural attractions and cultural resources. This promotes eco-tourism that keeps conservation in mind. Economic benefits come along without messing things up (Ramaano & A.I., 2025). As more people turn to GIS for rural planning, it shows how these tools work as full platforms. They pull together environmental science and economics. Social development gets integrated into one solid framework. GIS acts as a decision support tool. It serves as a communication aid, too. This lets community leaders collaborate with national agencies. They build sustainable futures for rural areas

4.2 Opportunities and Applications of GIS in Rural Development

The use of GIS in rural areas opens up all its possibilities through combining various data sources into one analytical framework. As per Scott and Rajabifard (2017), fairly a large number of development challenges have a geographical location, and through geospatial data, the needed information for tracking, forecasting, and controlling the way to the United Nations Sustainable Development Goals (SDGs) is provided. By using GIS, urban planners get the opportunity to look at various maps with their features such as social grouping, medical services, and environmental aspects and see how poverty, use of natural resources, and access to important services are related to each other. The spatial knowledge gained here will not only support policy making but also help in distributing the resources fairly. Participatory GIS (PGIS) has been hailed as the change-maker in people getting access to spatial data. It showed how quite remote the people in Sri Lanka were in getting the spatial data directly through their smartphones or by the use of satellites. In the mapping of the land use, roads, and connection of services, the participation of villagers was prominent while the youth community was engaged in the collection and digitization of data. This not only gave the local people power but at the same time led to the precision of village development plans.

PGIS is the connection between the planning from the top and the reality at the grassroots. The former is characterized by a lack of local people's demands; thus, the wrong kind of interventions are made. On the other hand, when the communities are allowed to participate in the mapping of the places, they start to discuss their needs and also co-design the development projects. This is how GIS makes the governance processes transparent, inclusive, and accountable. Besides, GIS is also the main tool in managing natural resources, reducing disaster risks, and adapting to climate change by providing reliable information for decision-making. More to that, GIS is also quite helpful in the monitoring and evaluation of such situations, as it shows the changes through time by creating dynamic dashboards and spatial indicators. In essence, GIS transforms data into actionable knowledge, bridging the gap between technological innovation and rural empowerment. Its participatory use ensures that rural development aligns with the needs and aspirations of the people it serves.

4.3 Building Blocks of GIS Effectiveness

Geographic Information Systems (GIS) play a vital role in effective rural planning, hinging on five interconnected key components: hardware, software, data, personnel, and methods. The seamless integration of these elements is essential for the success of the Decision-Making Support Framework that GIS offers.

The Five Key Components

- **Hardware:** This foundational element consists of physical tools such as computers, servers, GPS receivers, and remote sensors. These devices are crucial for the initial stage of gathering geospatial data.

- **Software:** A range of sophisticated tools, including ArcGIS, QGIS, and ERDAS Imagine, process the collected data. These applications facilitate key analytical functions such as overlay analysis, spatial analysis, statistical modeling, and data visualization.
- **Data (The Most Critical Element):** Data is of utmost importance, encompassing both spatial (location) and attribute (descriptive) information. This data is obtained from sources like remote sensing, satellite imagery, and field surveys. As Scott and Rajabifard (2017) emphasize, high-quality, detailed data is essential for accurately monitoring progress toward the Sustainable Development Goals (SDGs).
- **Personnel:** Even the most advanced technology requires human expertise. The involvement of knowledgeable individuals—such as planners, GIS analysts, and community members—transforms raw information into actionable strategic insights.
- **Methods:** Established methodologies are crucial for ensuring the integrity and accuracy of spatial analyses, thereby validating the resulting planning insights.

Integration for Effective Planning

When these components function cohesively, GIS simplifies complex environmental, social, and economic data. It transforms fragmented information into clear, interpretable spatial models. This powerful integration delivers the essential strategic knowledge required for effective, inclusive, and well-informed rural planning.

4.4 Challenges in GIS Implementation

GIS technology holds considerable promise for bringing changes to rural areas. Yet, efforts to adopt it continue to encounter several obstacles. Among these, infrastructure shortcomings stand out prominently. These include the lack of reliable computing equipment, inconsistent access to electricity, and limited internet connectivity (Hargreaves & Watmough, 2021). Moreover, many rural government institutions also face a problem of insufficient data, as most of the data available is either old or inconsistent. The limitation of human resources further complicates the issue. The lack of professional GIS staff in the fields of modeling, analysis, and management of data makes the operation of the system less effective (Fallah & Karami, 2023). Besides, according to Scott and Rajabifard (2017), the fragmentation of institutions and lack of interoperability among them have resulted in non-integration of geospatial data into the policy-making processes. Another category of challenges is socio-technical factors that affect the accessibility of GIS. Stakeholder low awareness, digital illiteracy, and unequal access to technology result in low participation from the community. Such disparities are more evident in the developing regions, where socio-economic inequalities are often like those in the digital sphere.

- **Availability, quality, and heterogeneity of data.**

The planning of rural areas usually involves the use of different data types, which include cadastral maps, satellite images, surveys, and community-collected data. These data vary significantly in terms of scale, format, projection, and time; missing, outdated, or inconsistent attribute information lowers the trustworthiness of analyses and may lead to false planning outputs (Wu, Gan, Chao, & Yu, 2024; Leeonis et al., 2024).

- **Expensive acquisition and maintenance.**

High-resolution imagery collection, commercial basemap purchase, server maintenance and GIS software licensing are hefty financial burdens for local governments and rural agencies. Limited budgets result in the GIS service being only partially or occasionally implemented rather than being a sustained, institutionalized practice (Leeonis et al., 2024; Ye, 2018).

- **Limited technical capacity and human resources**.

The majority of rural planning departments do not have staff with backgrounds in GIS, remote sensing, spatial analysis, database management, and GeoAI. The absence of those skills in-house limits the use of advanced techniques (like suitability mapping, spatio-temporal forecasting) and leads to greater reliance on external consultants, which in turn increases the cost and diminishes local ownership (Ceccato, 2000; Ye, 2018).

- **Institutional fragmentation and poor coordination**.

Data required for rural planning are often held in isolation by different departments (land records, agriculture, water, revenue). The lack of coordination among agencies, unclear data-sharing policies, and the absence of a centralized geospatial data infrastructure lead to a delay in the flow of information and the duplication of work (Leeonis et al., 2024).

- **Privacy, governance, and ethical concerns.**

Integration of high-resolution location data, household surveys, and mobile traces can expose personal and sensitive information. In rural contexts — where property disputes, communal land rights, and vulnerable populations are common — poor data governance can cause harm if privacy and consent are not addressed (Wu et al., 2024).

- **Big and heterogeneous geospatial data bring about technical obstacles.**

The advent of geospatial big data (satellite archives, drone images, IoT sensors) imposes the challenges of storage, indexing, retrieval, and processing. Ordinary GIS setups frequently prove to be inefficient for dealing with data at the petabyte level, especially time series or streaming sensor data, unless specialized surroundings are applied (Wu et al., 2024).

- **Spatial uncertainty and analytical weaknesses.**

Spatial uncertainty and analytical weaknesses characterize the case of rural analyses. The Uncertain Geographic Context Problem (UGCoP) and the Modifiable Areal Unit Problem (MAUP) are limitations that hinder rural analyses from getting the right picture (Ceccato, 2000; Wu et al., 2024). The wrong choice of aggregation, scale differences, and positional errors can all lead to biased output of resource allocation, vulnerability mapping, and land-use suitability that are relied on for policy decisions.

- **Digital divide and accessibility issues.**

Poor internet connectivity, lack of affordable devices, and low digital literacy characterizing many rural areas act as factors that limit the acceptability and participatory potential of GIS-based tools used in planning (Ye, 2018). This digital divide inadvertently keeps away the very stakeholders that rural planning is meant to empower.

- **Sustainability and long-term maintenance issues.**

GIS pilot projects are prone to getting stuck after the initial funding dries up, as no long-term operational or maintenance plan (for software updates, data refreshes, and staff training) is defined (Leeonis et al., 2024). The benefits of GIS are diminished over time when there is no sustained funding and institutional ownership.

- **Policy and legal loopholes for land information modernization.**

The existence of outdated land records, unsettled titling regimes, and absent legal frameworks for digital records are some of the factors that have rendered GIS less effective in land-use planning and dispute resolution. Legal and procedural reforms are often required simultaneously with technical modernisation for (Leeonis et al., 2024) effectiveness .

4.5 Strategic. Practical Solutions for Embedding GIS within Rural Development

The knotty obstacles that jam the blending of Geospatial Information Systems (GIS) with rural development call for a far-reaching strategic roadmap. Such a roadmap ought to amplify expertise, marshal state-of-the-art technology, reengineer policy structures and secure participation that genuinely embraces every voice.

1. Ramping up skill chops while deepening the tech know-how.

Bridging the gap in understanding is vital for the long-term success of any GIS rollout.

- Instruction and Learning: Roll out tailored workshops aimed at players—urban planners, students and community leaders—to sharpen their technical grasp and nudge them toward embracing GIS tools.
- Early Involvement: Turn to strategies such as Participatory GIS (PGIS), a technique that research has demonstrated can slash costs and nurture a sense of stewardship when school-aged children and youth are brought into the data-collection process (Participatory GIS, for Rural Development, 2008).

2. Overhauling the Tech Landscape and Widening Its Accessibility

- Embracing scalable tech cuts costs. Widens accessibility.
- Open-Source Utilisation: Tap into community-crafted GIS platforms—QGIS being an example—to trim expenses and sidestep the overhead of proprietary software.
- Cloud-Based Solutions: Deploy cloud-hosted GIS tools that let agencies exchange data instantly, speeding up data-driven reactions to emerging challenges and fostering planning.

3. Policy, governance and standardisation

- Revamping the rulebook is indispensable for sparking effort and unifying data practices throughout a patchwork of jurisdictions.
- National Spatial Data Infrastructure (SDI): Rolling out an SDI is essential for interoperability and for keeping standards in line (Scott & Rajabifard, 2017).
- Harmonisation and Quality: SDIs pull national data systems into alignment, boosting data quality, reinforcing governance and enhancing transparency.

4. Promote participatory, inclusive planning and collaborative effort.

Integrating community knowledge into development planning is crucial, for guaranteeing that initiatives stay both equitable and precisely targeted.

- Participatory GIS (PGIS): Weaving PGIS into the policy-making process widens participation. Research shows that when communities map their spaces, it gives a platform to the underserved, helping to shape initiatives that are both socially equitable and geographically precise (Afroza & Alam 2002; Ranasinghe, 2008).
- Web-GIS, for Transparency: Roll out info systems that naturally blend spatial and non-spatial data, boosting transparency and making information easier to access, as demonstrated by village-level models (in Sri Lanka).
- Strategic Partnerships: Cultivate public-private collaborations while actively involving universities and NGOs. This joint effort greatly lifts the precision and long-term sustainability of data.

These strategic moves not only patch up deficiencies but also propel GIS toward becoming a genuinely collaborative, inclusive platform for rural transformation.

6. CONCLUSION

Geographic Information Systems (GIS) represent a significant and transformative shift towards evidence-based governance and the stewardship of rural areas and practices. Beyond merely enhancing the effectiveness of development programs and policies, the integration of GIS tools through data-driven planning systems fosters ecological and social sustainability in rural regions. Previous research has demonstrated the utility of GIS in various applications, including disaster relief, water management, land-use planning, and agricultural mapping. It is anticipated that GIS will continue to serve as a crucial catalyst in transforming rural landscapes into resilient, self-sustaining ecosystems as countries strive to meet the Sustainable Development Goals (SDGs) established by the United Nations. However, despite the considerable potential of GIS technology to transform rural areas, its adoption faces several challenges. Among the primary obstacles are infrastructure limitations, including inadequate computing systems, unreliable power supplies, and poor internet connectivity (Hargreaves & Watmough, 2021).

Moreover, many rural government institutions also face a problem of insufficient data, as most of the data available is either old or inconsistent. The limitation of human resources further complicates the issue. The lack of professional GIS staff in the fields of modeling, analysis, and management of data makes the operation of the system less effective (Fallah & Karami, 2023). Besides, according to Scott and Rajabifard (2017), the fragmentation of institutions and lack of interoperability among them have resulted in non-integration of geospatial data into the policy-making processes. Another category of challenges is socio-technical factors that affect the accessibility of GIS. Stakeholder low awareness, digital illiteracy, and unequal access to technology result in low participation from the community. Such disparities are more evident in the developing regions, where socio-economic inequalities are often like those in the digital sphere.

7. LIMITATIONS AND SUGGESTIONS

The present study, while it makes a valuable contribution to the field of GIS and sustainable rural development, suffers from several inherent limitations. Firstly, the study is theoretical in nature, based on the existing research in the field of GIS and rural planning. Concerning this, empirical research is needed to get deeper insights into the practical applications of GIS in sustainable rural development. Secondly, due to the time constraints, the study focuses only on a few aspects of GIS integration, thus leaving a scope for further examination.

REFERENCES

Abdalla, R. (2016). *Introduction to geospatial information and communication technology (GeoICT)*. Springer.

Akgün, A. A., Baycan, T., & Nijkamp, P. (2015). Rethinking on sustainable rural development. *European Planning Studies*, *23*(4), 678–692.

Ashley, C., & Maxwell, S. (2001). Rethinking rural development. *Development Policy Review*, *19*(4), 395–425.

Bill, R., Blankenbach, J., Breunig, M., Haunert, J. H., Heipke, C., Herle, S., & Werner, M. (2022). Geospatial information research: State of the art, case studies and future perspectives. *PFG–Journal of Photogrammetry. Remote Sensing and Geoinformation Science*, *90*(4), 349–389. PMID: 36158628

Blasch, E. P., Pellechia, M., Deignan, P. B., Palaniappan, K., Dockstader, S. L., & Seetharaman, G. (2011, July). Contemporary concerns in geographical/geospatial information systems (GIS) processing. In *Proceedings of the 2011 IEEE National Aerospace and Electronics Conference (NAECON)* (pp. 183-190). IEEE.

Bruckmeier, K., & Tovey, H. (2008). Knowledge in sustainable rural development: From forms of knowledge to knowledge processes. *Sociologia Ruralis*, *48*(3), 313–329.

Ceccato, V. A., & Snickars, F. (2000). Adapting GIS technology to the needs of local planning. *Environment and Planning. B, Planning & Design*, *27*(6), 923–937.

Dangermond, J., & Goodchild, M. F. (2020). Building geospatial infrastructure. *Geo-Spatial Information Science*, *23*(1), 1–9.

. Deichmann, U., & Wood, S. (2001). GIS, GPS, and remote sensing.

Dwivedi, S., Singh, A. K., Gupta, A. K., & Chand, S. (2024). Introduction to Sustainable Development and Geospatial Technologies. In *Sustainable Development and Geospatial Technology: Volume 1: Foundations and Innovations* (pp. 1-16). Cham: Springer Nature Switzerland.

Fallah, M., Karami, T., & Kojouri, M. K. (2023). The Art Of Human Participation Against Artificial Intelligence With Geospatial Information System In The Tourism Industry. *Geo Journal of Tourism and Geosites*, *49*(3).

Hargreaves, P. K., & Watmough, G. R. (2021). Satellite Earth observation to support sustainable rural development. *International Journal of Applied Earth Observation and Geoinformation*, *103*, 102466.

Hossain, M., Park, S., & Shahid, S. (2023). Frugal innovation for sustainable rural development. *Technological Forecasting and Social Change*, *193*, 122662.

Karunathilake, E. M. B. M., Le, A. T., Heo, S., Chung, Y. S., & Mansoor, S. (2023). The path to smart farming: Innovations and opportunities in precision agriculture. *Agriculture*, *13*(8), 1593.

Kurowska, K., Marks-Bielska, R., Bielski, S., Aleknavičius, A., & Kowalczyk, C. (2020). Geographic information systems and the sustainable development of rural areas. *Land (Basel)*, *10*(1), 6.

Leeonis, A. N., Ahmed, M. F., Mokhtar, M. B., Lim, C. K., & Halder, B. (2024). Challenges of Using a Geographic Information System (GIS) in Managing Flash Floods in Shah Alam, Malaysia. *Sustainability*, *16*(17), 7528.

. Lo, C. P. and Yeung, A. K. W. (2007) Concepts

Mandal, D., & Ghosh, S. K. (2000). Precision farming–The emerging concept of agriculture for today and tomorrow. *Current Science*, *79*(12), 1644–1647.

Meeks, W. L., & Dasgupta, S. (2004). Geospatial information utility: An estimation of the relevance of geospatial information to users. *Decision Support Systems*, *38*(1), 47–63.

. Mihai, F. C., & Iatu, C. (2020). Sustainable rural development under Agenda 2030. *Sustainability Assessment at the 21st century*, 9-18.

Pandey, P. C., & Pandey, M. (2023). Highlighting the role of agriculture and geospatial technology in food security and sustainable development goals. *Sustainable Development*, *31*(5), 3175–3195.

Ramaano, A. I. (2025). The essence of geographic information systems (GIS) in sustainable tourism, public leadership and inclusive community participation in remote-African rural societies. *Journal of Responsible Production and Consumption*, *2*(1), 25–47.

Schultink, G. (2000). Critical environmental indicators: Performance indices and assessment models for sustainable rural development planning. *Ecological Modelling*, *130*(1-3), 47–58.

Scott, G., & Rajabifard, A. (2017). Sustainable development and geospatial information: A strategic framework for integrating a global policy agenda into national geospatial capabilities. *Geo-Spatial Information Science*, *20*(2), 59–76.

Shahid, M. K., Khin, A. A., Seong, L. C., Shahbaz, M., & Ahmad, F. (2024). Mapping the relationship of research and development expenditures and economic growth through bibliometric analysis: A theoretical perspective. *Journal of the Knowledge Economy*, *15*(4), 17529–17555.

Wu, J., Gan, W., Chao, H. C., & Yu, P. S. (2024). Geospatial big data: Survey and challenges. *IEEE Journal of Selected Topics in Applied Earth Observations and Remote Sensing*.

Ye, H., Brown, M., & Harding, J. (2018). GIS for all: exploring the barriers and opportunities for underexploited GIS applications.

Chapter 4
Assessing Kolkata's Urban Sprawl:
Quantifying Agglomeration Expansion Using Geospatial Technologies

Jitajitya Barh
Amity Institute of Geoinformatics and Remote Sensing, Amity University, Noida, India

K. Dhanaraj
https://orcid.org/0000-0002-9121-7314
Amity Institute of Geoinformatics and Remote Sensing, Amity University, Noida, India

Roshan Das
Department of Geography, Adamas University, Kolkata, India

Sinchan Banerjee
https://orcid.org/0009-0003-1859-5401
Department of Geography, Adamas University, Kolkata, India

Soumak Roy
Department of Geography, Adamas University, Kolkata, India

Prangya Rath
https://orcid.org/0000-0002-6764-1625
Amity Institute of Environmental Sciences, Amity University, Noida, India

Laxmi Kant Bhardwaj
https://orcid.org/0000-0001-7518-4199
Amity Institute of Environmental Toxicology, Safety, and Management, Amity University, Noida, India

ABSTRACT

Urban sprawl has emerged as a critical challenge in the context of developing nations. The KMDA region has undergone a remarkable spatial transformation over the past three decades. This study aims to assess, quantify, and analyze the spatial

DOI: 10.4018/979-8-3373-6320-2.ch004

and temporal patterns of urban sprawl in the KMDA region from 1991 to 2021 using multi-temporal satellite data and advanced geospatial techniques. The analysis employs Landsat Thematic Mapper and OLI-TIRS imagery from 1991, 2001, 2011, and 2021 to derive LULC classifications using the supervised Maximum Likelihood method. To evaluate the spatial dispersion of urban expansion, Shannon's entropy model was employed, revealing entropy values ranging from 2.33 in 1991 to 2.95 in 2011, signifying increasing urban sprawl and spatial diffusion. The findings reveal an outward migration of the urban core from ~ 6 km in 1991 to nearly 14 km in 2021. This study will contribute to a deeper understanding of urbanization processes in rapidly growing South Asian cities.

1. INTRODUCTION

The Earth's surface consists of fundamental elements, Land Use and Land Cover (LULC), that play a crucial role in the formation of the environment and influence human activity (Mir et al., 2025). Land management strategies, environmental conservation practices, and sustainability activities all rely on the understanding and dynamics of LULC. Specifically, the word 'land use,' according to the Intergovernmental Panel on Climate Change (IPCC), outlines the uses and activities carried out on the land by humans, such as forestry, mining, urbanization, transportation of goods and people, recreation, and conservation. In doing so, LULC demonstrates the depiction of societal demands, economic activities, and cultural and traditional factors concerning human society and the natural environment. On the other hand, land cover presents the characteristics of the materials that cover the Earth's surface. It comprises factors such as man-made structures, including roads, buildings, and industrial places, and natural structures, such as wetlands, forests, grasslands, deserts, and other earth bodies (KEBEDE, 2015).

The study's goals are to ascertain each land use/cover element's changing trajectory, stability, and interaction with other elements, as well as to identify those elements that exhibit similar changing features. Making best use of land requires having knowledge about the current land cover and use throughout regional areas. Because of the demands of a more populous people and the forces of nature sculpting the landscape, it is also critical to be able to track the shifting of land use. The terrain is constantly changing as a result of numerous natural and human-made processes. Research concentrating on the peri-urban and rural neighborhoods that also face comparable issues is lacking. As a result, it is essential to keep an eye on these problems and take necessary action by putting policies in place. The current study chooses the densely populated and quickly growing KMDA and its surrounding district area of West Bengal as the study location in order to address this issue.

The majority of previous research focuses on examining the pattern of LULC modifications using modeling techniques. Liping et al. (2018) predicted Jiangle's likely future land use trends in China using a Cellular automata Markov analysis. Conventional strategies for identifying changes in land use and cover are costly, imprecise, and only successful in small areas. High-resolution satellite imagery, enhanced image processing, and GIS led to more consistent monitoring and modeling of land use/cover trends, with extensive applicability for spatio-temporal study. Land use/cover change detection involves studying an object or environment over time to detect changes (Xiaomei et al., 1999).

The Entropy method, which combines remote sensing and GIS, is a popular metric for measuring urban sprawl (Yeh and Li, 2001). Sprawl can be quantified in relative and absolute scales (Bhatta et al., 2010). Researchers have developed their own assumptions while establishing this threshold, which are unclear to scientists (Bhatta et al., 2010). Quantifying sprawl as a pattern or process poses a significant challenge (Bhatta et al., 2010). According to Wilson et al. (2003), it is difficult to quantify and simulate urban sprawl without a universal definition.

Dhanaraj and Angadi (2022) estimate the urban form of demarcation as urban core, inner core, suburban, and urban fringe, depending on the density of the study area. The rapid growth of a city's built-up area is primarily due to intense infill growth in its core range, as evidenced by the Shannon entropy value, indicating compact growth patterns. The research on compact development schemes in the urban core of KMDA reveals the profound effects of urbanization, the lack of efficient spatial planning, and the changes in the socioeconomic structure, which are the main factors that reshape metropolitan regions in developing countries. As a consequence of unregulated urban expansion, the city of Kolkata Metropolitan Region (KMR) has, during the last few decades, shown the features of such a situation - growth has been concentrated in the core with a vast sprawl of the peri-urban area (Bhatta et al., 2010). This situation reflects not only the centripetal forces of economic opportunity but also the centrifugal forces of population displacement, industrial decentralization, and infrastructural strain. The morphological change thus resulting is one that combines a heavily populated central business district with increasingly fragmented suburban and rural areas that are left behind, thereby generating spatial heterogeneity in terms of land use and (Sang et al., 2011).

Such patterns of expansion are frequently the effects of a rise in population and changes in the economy. The city spread in the case of KMDA has thus mostly been a result of the growth of the informal settlements, the establishment of the industrial clusters on the outskirts, and the creation of the transit corridors along the Hooghly River. The development of the linear settlements around the Eastern Metropolitan Bypass and Kona Expressway is a clear indication of how the expansion of the transport infrastructure leads to changes in land use, both in terms of space and time (Halder et al., 2025). The conversion of agricultural land into residential and commercial areas has been going at a very fast rate due to the market, driven by land speculation and the shortage of proper zoning regulations. Consequently, the farmlands of North and South 24 Parganas have experienced major changes, with rice fields and wetlands giving way to mixed built-up areas (Das and Angadi, 2022). The change here is not only that of the physical landscape but also a huge socio-economic transition of the community from agricultural to service-based occupations.

By referring to a broad theoretical framework, the phenomenon of urban sprawl in KMDA can be considered as "ribbon development", which is a spatial typology widely discussed in the context of South Asian metropolitan regions (Sudhira et al., 2004). The development of such a structure happens when the extension of the city follows the transport means that connect the different parts instead of a growth of a compact, concentric form. Certainly, this kind of growth results in easier access to transportation; however, it also causes environmental pollution through the degradation of the green cover, the loss of the wetlands, and an increase in air and water pollution. Furthermore, remote sensing data studies from the years 1990 until 2020 have shown significant changes around the eastern and southern parts of the city of Kolkata, such as a considerable increase in the area covered by hardened surfaces, along with a sharp decrease in the vegetation indices (Mahata et al., 2024). The disappearance of the natural buffers like the East Kolkata Wetlands, a Ramsar site, is only one example explaining the ecological prices that are paid for the unsustainable LULC changes.

By integrating remote sensing with GIS methods, we have considerably improved the means to measure and demonstrate the changes caused by such phenomena. The usage of satellite datasets like Landsat Thematic Mapper (TM), Enhanced Thematic Mapper (ETM) +, and Operation Land Imager (OLI) in conjunction with the indices such as Normalized Difference Built-up Index (NDBI) and Normalized Difference Vegetation Index (NDVI) enables the researchers to pinpoint spatio-temporal patterns of urban expansion with a great accuracy level (Chanu et al., 2021). The study of changes over time through KMDA images reveals that the artificial surfaces in the Kolkata Metropolitan area have increased by almost two times between 1990 and 2020, mainly due to the reduction of vegetation and water bodies." Furthermore, the use of Shannon's entropy in this case gives a precise account of the extent of the spread, showing that there is an increasing spatial randomness of the distribution of the built-up areas, especially in the outskirts (Lata et al., 2001). These results support the idea that the development of Kolkata is going away from the compact type towards the dispersed one, and therefore, the rise of secondary urban centers in Rajarhat, Baruipur, and Serampore (Sahana et al., 2018).

The Cellular Automata, Markov modeling technique, delineates with greater detail changes in the LULC patterns that will happen in the future. As the Cellular Automata, the Markov model involves spatial transition probabilities with a rule, based on neighborhood interactions, which can depict both stochastic and deterministic aspects of urban growth (Sang et al., 2011). Such models in KMDA have illustrated the scenarios that the urban areas will continue to get dense along the highways and rail lines, where it will be more difficult to find open spaces. And also, the next urban sprawl is expected to be in the agricultural, based, and wetland ecosystems of the peri-urban areas. The models' predictive capabilities are therefore an important decision, a support tool for planners and policymakers, who thus recognize the necessity of regulated urban infill and the conservation of green areas (Dutta and Das, 2019).

Similar patterns are largely found in the other Indian metropolitan areas as well. For example, the National Capital Region (NCR) of Delhi has undergone a rapid peri, peri-urbanization that is mainly a consequence of the industrial decentralization and real estate expansion in the areas of Gurgaon, Noida, and Ghaziabad (Acharya et al., 2017). As in the case of KMDA, high entropy values are typical of the NCR, which is a sign of fragmented and dispersed growth (Jat et al., 2008). The Mumbai Metropolitan Region, limited by the coastal geography, shows a different kind of vertical densification of the area with the absence of a lateral sprawl; however, the increase in demand for land and the depletion of ecosystem services are at a similar level (Kamini et al., 2006). Chennai and Hyderabad also have the same tendencies that have been described in the text above, district expansion beyond municipal boundaries, which has led to rural, urban interfaces that are less clear and therefore more difficult to manage (Chettry, 2022). These comparisons reveal the commonality of the urban sprawl phenomena in Indian megacities that have been described as poorly planned, unregulated real estate markets, and infrastructural lag in the studies.

Worldwide, the stadia visible in KMDA echo the urban changes visible in quickly developing cities of Shanghai, Lagos, and São Paulo (Zhao et al., 2017). After reform, Shanghai saw the massive move of industries out of the city along with the liberalization of the economy, which led to the huge change of agricultural land to industrial and residential land, very much similar to the change along the edges of Kolkata (Wu, 2016). Lagos, Nigeria, is another instance of an area that has grown beyond control, where the establishment of informal settlements and poor governance have been the main contributors to the disorder of space, and the area is becoming more vulnerable to the degradation of nature (Parnell and Walawege, 2011). The periphery of São Paulo is a reflection of that of Kolkata, with socio-spatial disparities being the major issue as the peripheral districts become the ones that absorb the urban poor, while the central areas undergo gentrification and get the infrastructural deposits (Lukas and Reis, 2022). The local parallels reveal that, as urban sprawl differs from one place to another depending on the local conditions and the type of governance, the fundamental causes of it, demographic increase, economic transition, and absence of proper policies, are almost the same anywhere.

Moreover, the unregulated changes in LULC of urban areas such as KMDA have caused a decline in the quality of the environment, which is one of the main consequences. The replacement of permeable surfaces with concrete constructions not only intensifies the city or urban heat island (UHI) effect but also alters local hydrology and increases the risk of floods (Vujovic et al., 2021). The research shows that the average surface temperature of Kolkata has risen substantially in the areas that are densely built, especially in the central and eastern parts, which is likewise related to the reduction of the vegetation cover (Chatterjee and Majumdar, 2022). The reduced capacity of wetlands and open spaces to absorb rainfall has been the cause of flooding in urban areas, which has made the flood situation a regular occurrence in the monsoon season. Besides that, the disappearance of trees results in less carbon sequestration potential and worsened air quality, which, in turn, affects human health (Wang and Cao, 2011).

Land Use and Land Cover (LULC) changes are closely linked with the alterations of people's livelihoods and the socio-economic conditions of the areas (Desalegn et al., 2014). Besides the displacement of rural and peri-urban communities due to land acquisition for urban expansion, which ditches their socio-economic way of life, the uprooting causes their socio-economic marginalization as a consequence. Informal housing colonies have been spreading like a fire has been raging rapidly in the outskirts of the KMDA region, where the residents are devoid of the proper provision of water, sanitation, and transportation facilities (Adiba, 2023). The spatial inequality perpetuates the socio-environmental vulnerability cycle as the marginalized populations become the most affected by the environmental hazards, such as flooding and pollution (Dutta and Das, 2019). Besides that, the central city only facilitates the congestion and affordability problems to become worse by the concentration of economic opportunities there, thus indicating the pressing demand for balanced regional development. Moreover, the problem of overcrowding and high cost of living is only made worse by the concentration of economic opportunities in the central city, which therefore points to the urgent need for balanced regional development (Jat et al., 2008).

The use of geospatial technologies in the monitoring of LULC changes has helped to improve the spatial accuracy of the data and has also facilitated the multi-temporal policy assessments. The combination of remote sensing with socio-economic data makes it possible to have a complete diagnosis of an urban area that considers physical, demographic, and infrastructural variables as well (Sahana et al., 2018). The use of spatial metrics like patch density, edge density, and landscape shape index helps in quantifying the level of fragmentation more accurately, thus giving a more nuanced understanding of the morphological evolution of cities (Herold et al., 2003). These methods are not limited to land cover mapping only; they provide a deeper understanding of the processes and patterns that underlie urban growth.

One significant change in LULC research will be the incorporation of machine learning (ML) and artificial intelligence (AI) to not only detect changes but also to predict future scenarios. For instance, the Random Forest (RF), Support Vector Machines (SVMs), and Artificial Neural Networks (ANNs) algorithms are, reputedly, the most advanced, cutting-edge tools that are being leveraged more and more to analyze complex urban landscapes and to make predictions about their changes in the future (Halder et al., 2025). These models enhance performance as they identify complicated, non-linear spectral data relationships for different land cover classes. An AI-driven tool might be a huge urban planning asset, which is able to simulate different development scenarios for the KMDA area, from business as usual, for example, growth, to green infrastructure, based on expansion. Such urban governance solutions fueled by data-driven decisions open up new avenues for planning the rapid urbanization that results in unplanned sprawl and environmental degradation (Chatterjee and Majumdar, 2022).

While improvements in methods have been made, policy frameworks still have the most influence in determining land use results. The management of land, however, is still very much affected by the splitting up of institutions and the presence of several authorities that the Kobe Metropolitan Development Plan and the National Urban Renewal Mission have not been able to deal with successfully (Bhatta et al., 2010). Metropolitan governance that is well coordinated is essential to be able to combine spatial planning with environmental protection. For example, aside from the urban development authorities, the pollution control board, and the local municipalities, inter-departmental collaboration is necessary for the conservation of the East Kolkata Wetlands. Different successful models from other areas, like Singapore's integrated land use planning or Curitiba's transit-oriented development, show how institutional stability can solve the problem of combining growth with sustainability (Hall, 2014).

Maintaining LULC change in check and under control is much in conformity with the United Nations Sustainable Development Goals (UNSDGs). This holds particularly true for Goal 11 ("Sustainable Cities and Communities") and Goal 15 ("Life on Land"). The experience of the Metropolitan Development Authority suggests that urban expansion will be sustainable only if economic progress is combined with ecological conservation and social inclusivity. The use of green belt zoning, urban greening, and wetland rehabilitation, among others, has been found to work in mitigating the adverse effects of the dispersed growth of urban settlements. In addition, the promotion of mixed, use and high, and high-density development within the respective urban areas may result in maximum utilization of land, while the transport emissions level will be reduced (Neuman, 2005). In addition, the use of spatial decision support systems in conjunction with Geographic Information Systems (GIS) may be an excellent aid for improving planning stakeholders' participation by facilitating their ability to analyze various scenarios.

At the same time, metropolitan cities like Delhi and Bengaluru are redesigning their city master plans by including environmental protection as one of the priorities through green infrastructure strategies and by giving a monetary value to the services provided by the ecosystem (Saxena et al., 2025). KMDA could take the cue from how Delhi is creating "green buffer zones" and Bengaluru is implementing "lake rejuvenation programs" (Dutta and Das, 2019). In fact, Shanghai's greenbelt and London's ecological redlining policies are two instances from different parts of the world, which show how policy instruments can be used efficiently to stop urban sprawl and, at the same time, protect the environment. Nevertheless, it is very important to change the aspect. For instance, the density zone of Kolkata, the old land tenure pattern, and the socio-economic mix of the area require that the methods be altered to fit the local conditions prevailing (Chettry, 2022).

Community programs, which are beyond the influence of policy and planning, can be the most important factor in the environmentally friendly management of land use. Local groups, citizen science projects, and community mapping activities have been recognized as effective instruments for developing local environmental awareness and for having a say in the management of the city (Das and Angadi, 2022). In KMDA, participatory GIS platforms can be used to document wetland encroachments, illegal constructions, and polluted areas, thus spreading the idea of a different sense of responsibility at various levels. The combination of such social and technological strategies is an example of the transition to more open and flexible urban management systems of the future (Das and Angadi, 2022).

The focus on technological tools and spatial modelling is great, but just as important is the acknowledgment of the socio-political aspects of LULC change. The development of land, the informal sectors, and the inequalities of governance are, in fact, the factors that impel the expansion of the city to a very large extent, whereas the predictions of technical models are overruled (Chatterjee and Majumdar, 2022). Therefore, subsequent investigation should combine interdisciplinary frameworks that not only integrate spatial science but also include social, economic, and political analysis. Such studies, over time, linking LULC change with demographic and policy variables, can uncover the underlying mechanisms, thus providing guidance for government decision-making based on the real facts (Chanu et al., 2021).

The KMDA case, indeed, teaches us a very crucial lesson that can be applied to other developing nations: uncontrolled urban expansion, most often the situation, will lead to immense threats to environmental sustainability, social justice, and economic vulnerability (Mahata et al., 2024). Thus, land administration must be founded on three pillars: robust spatial information systems, participatory government, and regulatory systems that can be implemented. With urban settlements increasing in size through the combined impact of globalization and climate change, it is extremely important that LULC monitoring is mainstreamed in city policy processes. Cities such as Kolkata will be able to survive only if they are successful in balancing growth with the conservation of their ecological as well as cultural landscape (Chettry, 2022).

2. MATERIALS AND METHODS

2.1. Study Area

In the eastern region of India, the Kolkata metropolitan area is served by the Kolkata Metropolitan Development Authority (KMDA) (Figure 1). The research area's principal physical location is the West Bengal state, which includes Kolkata and the metropolitan and peri-urban areas that surround it. KMDA, under its jurisdiction, there are Six Districts: Kolkata, Hooghly, Howrah, Nadia, North 24 Pargana, and South 24 Pargana, with a surface area of 1851 km^2 that extends from 22°17′ N to 23°03′ N and 88°06′ to 88°52′ E (Ray et al., 2023). Its infectious linear urban concentration is found on the Hooghly River's eastern and western banks. The agglomeration comprises 38 municipalities, 77 non-municipal urban towns, 16 outgrowths, 3 municipal corporations (Howrah, Kolkata, Chandan Nagar), and 445 rural villages. With an average population density of 7950 people/km^2, the agglomeration has 14.72 million people overall (Census of India, 2001). By 2025, the population is expected to reach 21.1 million, growing at a 1.8% annual rate (Census of India, 2001; KMDA, 2011). Kolkata is a developing country's expanding metropolis that has significant issues with traffic congestion, poverty, overcrowding, urban pollution, and socioeconomic issues. Under the administrative direction of the Government of West Bengal, the Urban Development Department oversees KMDA's statutory power functions.

Figure 1. Location map of Kolkata Metropolitan Development Authority (KMDA)

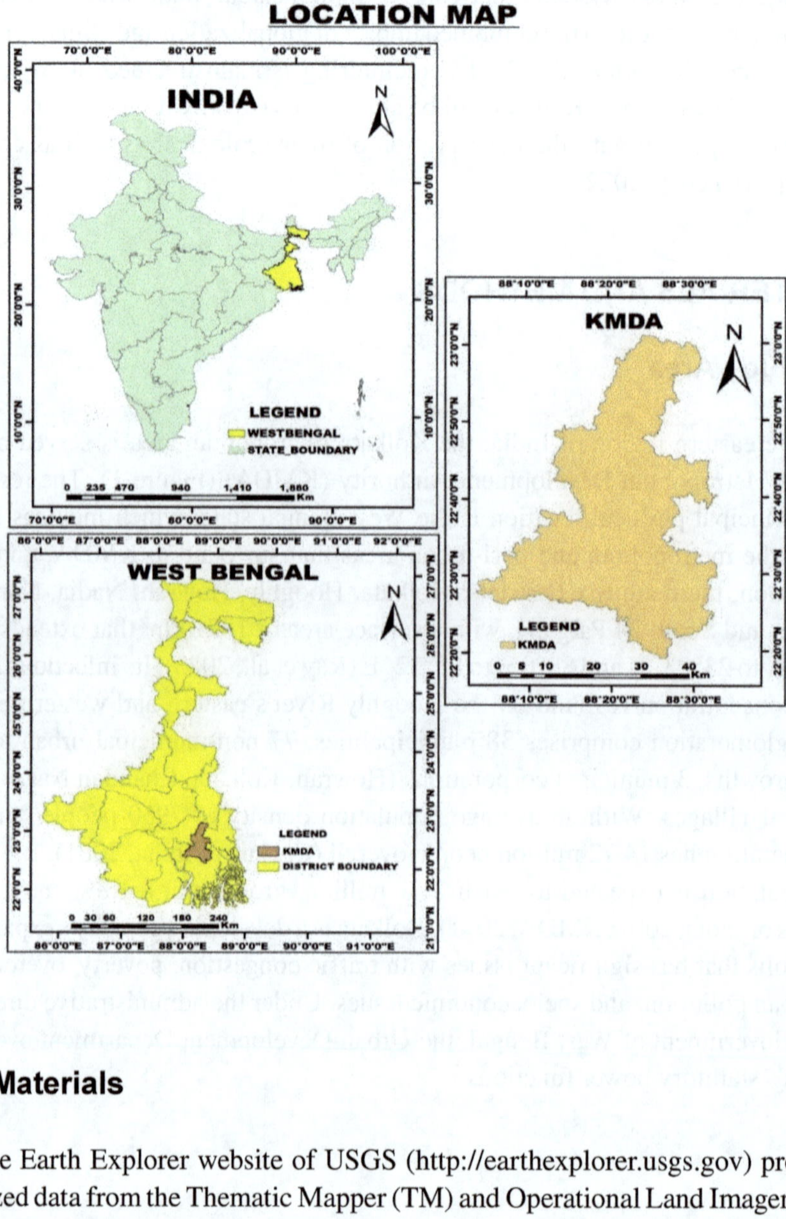

2.2. Materials

The Earth Explorer website of USGS (http://earthexplorer.usgs.gov) provided digitized data from the Thematic Mapper (TM) and Operational Land Imager (OLI) and Thermal Infrared Sensor (OLI-TIRS) sensors of the years 1991, 2001, 2011, and 2021 that covered the research region. The Landsat photos were chosen for this investigation due to their availability and medium spatial resolution to high resolution. Table 1 provides comprehensive details on the data. This data is referenced with UTM Zone 45N using WGS84 coordinate Datum.

Table 1. Datasets of sensor types

<table>
<tr><th>Year</th><th>Type of Sensor</th><th>Path-Row</th><th>Bands</th><th>Resolution</th></tr>
<tr><td>1991</td><td rowspan="3">Landsat Thematic Mapper (LTM)</td><td rowspan="4">138-44, 138-45</td><td rowspan="3">Seven (7)</td><td rowspan="4">30 m</td></tr>
<tr><td>2001</td></tr>
<tr><td>2011</td></tr>
<tr><td>2021</td><td>Landsat OLI_TIRS</td><td>Eleven (11)</td></tr>
</table>

2.3. Methods

2.3.1. Pre-Processing

We use spatial raster data from Earth Explorer for the years of 1991, 2001, 2011, 2021, Landsat 5 & 8 Thematic Mapper Path-Row 138-44/ 138-45, and spatial resolution of 30 m each. In order to highlight the most significant changes in the study area. We structured the time frames for the study as decadal intervals according to the decadal change period. To guarantee that the data are of the same quality and format, pre-processing procedures for both Landsat 8 & 5 data should be followed. The pre-processing processes for data consist of geometry adjustment, radiometric, and atmospheric correction. Before comparing the outcomes, it is crucial to normalize data from both Landsat 5 & 8 in order to highlight the variations in sensor properties. To make sure the data are on the same scale, we employed normalization techniques such as top-of-atmosphere reflectance. We then selected the suitable spectral bands for the study based on the particular needs. The acquired remote sensing data was integrated using the Earth Resource Data Analysis System (ERDAS) Imagine application and utilizing the layer stack option in the spectral toolbox, false color composite images were created. We cropped every image to isolate the research area based on the Area of Interest before moving on to the image classification step.

2.3.2. Image Classification

To produce the LULC classification, the False Color Composites (FCCs) with the best band combinations over the course of three decades were selected using the maximum likelihood method and the hard classification approach. This method is based on the probability density of pixels that belong to a certain class and presupposes that the probability densities of all the classes are identical. The maximum likelihood algorithm, like other supervised classification methods, finds spectral distance among mean vectors for each signature inside a cluster of pixels and the measurement vector for the candidate pixel. Using the classifier algorithm, the study area analyzed in this case study was divided into five LULC classes: Built-up area, waterbody, crop land, fallow land, and vegetation. Per-pixel signatures were assigned from corresponding satellite data based on the specific pixel value of reflectance. To ensure the efficacy and dependability of this approach, post-classification refinement was utilized, with the goal of enhancing classification accuracy and decreasing the occurrence of misclassification mistakes.

2.3.3. Accuracy Assessment

Using the error matrix and Kappa statistics, the classification accuracy of three decades' worth of studied and classed photos was evaluated in order to assess the quality of information gained from the data. In order to utilize the categorized outputs for change detection, an accuracy evaluation for each individual classification must be carried out. When gauging the veracity of LULC data or the caliber of accurate information gleaned from the data, accuracy assessment is crucial. A review of the accuracy of four satellite photos from 1991, 2001, 2011, and 2021 was conducted using from Google Earth service. Each categorized image was subjected to the overall accuracy and Kappa coefficient, and each LULC type was subjected to the producer and user accuracy. The accuracy of the extraction value and the real value was measured using the Kappa coefficient, overall accuracy, producer accuracy, and user accuracy. The following formulas can be used to determine these (Gwet, 2002; Viera and Garrett, 2005).

$$\text{Overall Accuracy} = (TP_{cd} / TP_r) \times 100 \qquad (1)$$

Where the total number of correctly classified diagonal cells (cd) is represented by TP_{cd}, while the total number of reference pixels (r) is represented by TP_p.

$$\text{User Accuracy} = (NCP_i/TCP_i) \times 100 \qquad (2)$$

Where the number of categories is denoted by $_i$; TCP_i is the total number of classified pixels in the ith category (the row total), and NCP_i is the number of correctly classified pixels in the ith category.

$$\text{Producer Accuracy} = (NCP_i/TP_{ri}) \times 100 \qquad (3)$$

where TPri, or the column total, is the total number of reference pixels in the ith category.

$$Kappa\ Coefficient = \left\{ \left(TPr \times \sum_{1}^{n} NCPi \right) - \left(\sum_{i}^{n} T\,\mathrm{Pr}i \times \sum_{i}^{n} TCPi \right) \right\} / TP^{2}{}_{r} - \left(\sum_{i}^{n} TP_{ri} \times \sum_{i}^{n} TCPi \right) \quad (4)$$

where the total number of categories is denoted by n. The Kappa coefficient result falls between 0.00 and 1.00. A value of 1.00 indicates perfect or strong agreement, whereas a value below 1.00 indicates less perfect or weaker agreement.

2.3.4. Built-Up Extraction

The dynamics of urban land use change are linked to fast-paced urbanization. Urban heat islands are just one of the many issues related to the urban environment that arise when natural land surfaces are replaced by artificial, impermeable built-up surfaces in urban clusters. It is crucial to employ satellite data and other spectral indices for remote sensing-based built-up area estimation in urban clusters. Therefore, the modification of previously produced spectral indices for the classification of built-up regions is the main emphasis of the current work. After discussing the principles and algorithms, the paper contrasts the results of various built-up area indices that were produced using various spatiotemporal satellite data. After pre-processing, LULC classification and accuracy assessment, we extracted only the Built-up area of each year, i.e, 1991, 2001, 2011, 2021, and calculated the area.

2.3.5. Change Detection

Finding out which land-use class is shifting to the other is crucial for change detection. Understanding the connections and interactions between natural and human occurrences is crucial for facilitating improved decision-making. This can be achieved through the timely and precise change detection of Earth's surface features. The two most often utilized techniques for change detection are principal component analysis and post-classification comparison.

2.3.6. Shannon Entropy

Shannon's entropy model has been used in a number of studies conducted in recent decades to examine and comprehend the equilibrium rate of the relative urban phenomena at the state, regional, and national levels (Shannon, 1948). On the other hand, there is a murmur that Shannon's entropy is a reliable indicator of the urban expansion process. Although this method has several limits, it is still preferable despite some annoyances. Furthermore, the results of Shannon's entropy and landscape metrics may show a conflicting relationship. In order to comprehend the effectiveness of the chosen measures, urban growth for a suggested zoning scheme was analyzed individually using Shannon's entropy computation.

The following formula is used to calculate the model's structure:

$$H_{n=} - \sum_{i=1}^{n} pi*loge(pi)$$

Where pi represents the variable's proportion (built-up land) in the ith zone.

$$Pi = xi / \left(\sum_{i=1}^{n} xi\right)$$

The variable's observed value in the ith zone is denoted by xi, while the total number of zones is represented by n. The range of the entropy value is 0 to log(n). The compact distribution is indicated by a number that is closer to zero, and the dispersed distribution is shown by a value that is near log(n). Expanding entropy levels are indicative of sprawl (Bhatta, 2012). According to Lacour and Feidt (2022), the term "spraying city" refers to an entropy value that exceeds the threshold, which is determined by measuring the halfway point of log(n). Scaling up the entropy value into a range from 0 to 1 can be achieved by measuring relative entropy. The spatial compactness or diffusion of the city is measured using Shannon's entropy value, which is computed for the KMDA for the years 1991, 2001, 2011, and 2021. Using ArcGIS's buffer feature, 25 concentric zones with a width of 2 km were made between the city's center and its border. The built-up layer obtained from the thematic map of classified land use and cover is then clipped to each buffer zone according to the years 1991, 2001, 2011, and 2021 in order to calculate Shannon's entropy value.

2.3.7. Built-Up Density and Built-Up Class

Based on a circular multiring buffer 2 km from the Central Business District (CBD), the built-up areas density curve for the years of 1991,2001,2011,2021 was created. The built-up density of a multiple concentric ring buffer, spaced two kilometers apart from the city's core business center, was determined by dividing the ring's total built-up area by its entire area. Water bodies found in each circular ring were removed during the density calculation process because it is unlikely that they will become part of a built-up region (Zhang et al., 2017). The built-up density is divided into 4 categories:>60%- urban core, 40-60%- inner core, 20-40% sub suburban, and <20% is urban fringe. Figure 2 shows the methodology.

Figure 2. Methodology

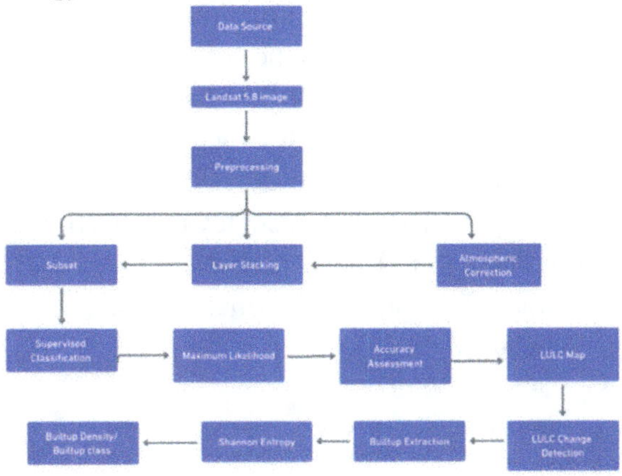

3. RESULT AND DISCUSSION

The results of the study area are discussed in 3 major parts, i.e, (1) LULC analysis and accuracy assessment of the years 1991, 2001, 2011, 2021, and (2) change detection analysis of different years, (3) Shannon entropy for estimating urban sprawl, (4) Built-Up density and Built-up Class classification.

3.1. Land Use and Land Cover (LULC)

Preprocessing of different years using Landsat 5 & 8 TM Path-Row 138-44/ 138-45 & spatial resolution of 30 m each. Analysis of LULC of the years 1991, 2001, 2011, and 2021 is done with supervised classification using the maximum likelihood method (Figure 3). Using the classifier algorithm, the study area analyzed in this case study was divided into five LULC classes: Built-up area, waterbody, crop land, fallow land, and vegetation. Per-pixel signatures were assigned from corresponding satellite data based on the specific pixel value of reflectance. Total area of the KMDA is 1851 Km^2. After classifying LULC of 1991, it finds the area of waterbody 97.1145 km^2, vegetation 561.2427 Km^2, built-up 139.3164 km^2, agricultural fallow land 605.5839 Km^2, and agricultural crop land 365.1156 km^2. For the year 2001 it finds waterbody 95.8752 km^2, vegetation 600.081 Km^2, built-up 301.8438 km^2, agricultural fallow land 93.159 Km^2, agricultural crop land 677.4138 km^2. For the year 2011, it finds waterbody 148.850 km^2, vegetation 344.237 Km^2, built-up 646.524 km^2, agricultural fallow land 219.277Km^2, agricultural crop land 412.535km^2. For

2021 it finds waterbody 153.955 km^2, vegetation 332.600 Km^2, Built-up 726.951 km^2, agricultural fallow land 80.671 Km^2, agricultural crop land 480.9117659 km^2. After analyzing the classified data, it seems the built-up area increased rapidly from 1991 to 2011, i.e, from 139.314Km^2 to 646.524 Km^2 (Table 2) (Figure 4). Using the error matrix and Kappa statistics, the classification accuracy of three decades' worth of studied and classed photos was evaluated in order to assess the quality of information gained from the data. The accuracy of the extraction value and the real value was measured using the Kappa coefficient, overall accuracy, producer accuracy, and user accuracy. The following formula can be used to determine these (Gwet, 2002; Viera and Garrett, 2005).

$$Kappa\ Coefficient = \left\{\left(TPr \times \sum_{1}^{n} NCPi\right) - \left(\sum_{i}^{n} T\,\mathrm{Pri} \times \sum_{i}^{n} TCPi\right)\right\} / TP^2{}_{r} - \left(\sum_{i}^{n} TP_{ri} \times \sum_{i}^{n} TCPi\right)$$

After applying the above formula, we find the accuracy of the classified LULC map, i.e, for the year 1991, overall accuracy 94.75% and overall Kappa 0.93. For the year 2001, overall accuracy was 87.75% and the overall kappa was 0.85. For the year 2011, overall accuracy was 92.25% and overall kappa was 0.90. For the year 2021 overall accuracy 85% and the overall kappa was 0.81.

Figure 3. Land Use and Land Cover (LULC) Map of Kolkata Metropolitan Development Authority (KMDA) of 1991, 2001, 2011, 2021

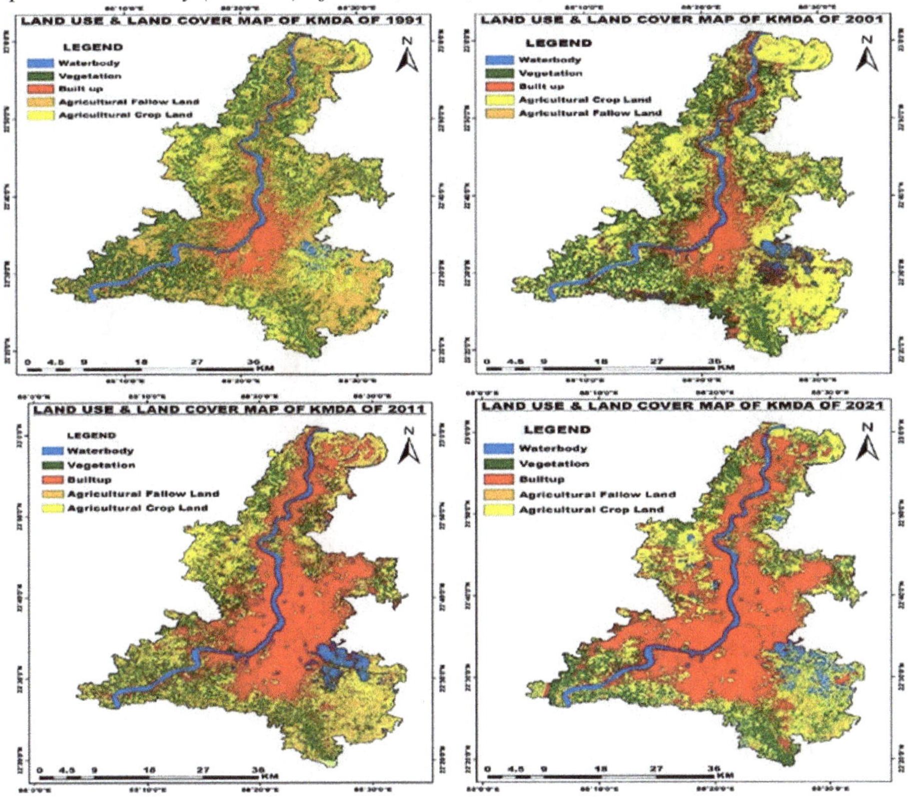

Table 2. Area calculation of Land Use and Land Cover (LULC) of 1991, 2001, 2011, 2021

Class	Area (km²) 1991	Area (km²) 2001	Area (km²) 2011	Area (km²) 2021
Waterbody	97.1145	95.8752	148.8507143	153.9550961
Vegetation	561.2427	600.0813	344.2371204	332.6008815
Built-up Area	139.3164	301.8438	646.5241552	726.9512245
Agricultural Fallow	605.5839	93.159	219.2770354	80.67129808
Agricultural Cropland	365.1156	677.4138	412.5352823	480.9117659

Graphical Representation of the Land Use and Land Cover (LULC) of Kolkata Metropolitan Development Authority (KMDA) for the Years 1991, 2001, 2011, 2021

Figure 4. Bar diagram showing Land Use and Land Cover (LULC) Area for the Years 1991, 2001, 2011, 2021

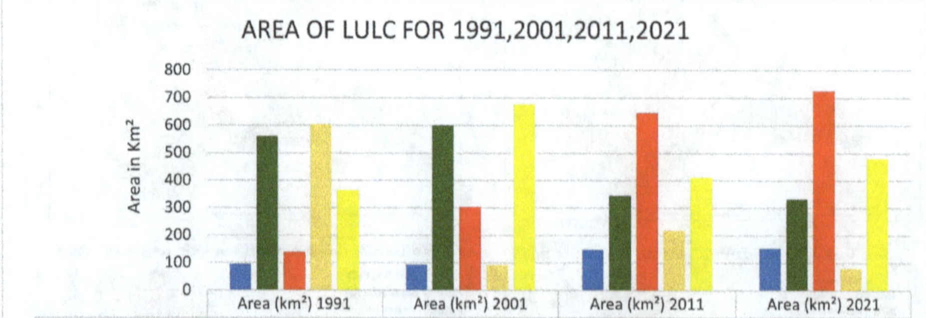

	Area (km²) 1991	Area (km²) 2001	Area (km²) 2011	Area (km²) 2021
Waterbody	97.1145	95.8752	148.8507143	153.9550961
Vegetation	561.2427	600.0813	344.2371204	332.6008815
Builtup Area	139.3164	301.8438	646.5241552	726.9512245
Agricultural Fallow	605.5839	93.159	219.2770354	80.67129808
Agricultural Cropland	365.1156	677.4138	412.5352823	480.9117659

3.2. Accuracy Assessment

Table 3 shows the accuracy assessment for the Years 1991, 2001, 2011, and 2021.

Table 3. Accuracy assessment for the years 1991, 2001, 2011, 2021

Year	Accuracy	Classes					Overall Accuracy	Overall Kappa
		Water bodies	Vegetation	Builtup	Agricultural Fallow	Agricultural Cropland		
1991	Reference Total	70	80	90	85	75	94.75	0.93
	Classified	70	80	90	85	75		
	Correctly Classified	58	65	75	70	72		
	User's Accuracy (%)	82.85	81.25	83.33	93	99		
	Producer's Accuracy (%)	92.00	95.00	94.00	96.00	97.00		
2001	Reference Total	65	85	82	93	75	87.75	0.85
	Classified	65	85	82	93	75		
	Correctly Classified	57	75	64	88	67		
	User's Accuracy (%)	87.69	88.23	78.04	95.00	89.00		
	Producer's Accuracy (%)	92.00	84.00	100.00	89.00	78		
2011	Reference Total	60	90	80	105	65	92.25	0.90
	Classified	60	90	80	105	65		
	Correctly Classified	53	85	74	95	62		
	User's Accuracy (%)	88.33	94.44	92.50	90.00	95		
	Producer's Accuracy (%)	91.00	92.00	90.00	91.00	97.00		
2021	Reference	70	80	90	85	75	85.00	0.81
	Classified	70	80	90	85	75		
	Correctly Classified	58	65	75	70	72		
	User's Accuracy (%)	82.85	81.25	83.33	82.00	96.00		
	Producer's Accuracy (%)	85.00	86.00	94.00	78.00	84		

1991 overall accuracy was 94.75% and overall Kappa 0.93. For the year 2001, overall accuracy was 87.75% and overall kappa was 0.85. For the year 2011, overall accuracy was 92.25% and overall kappa was 0.90. For the year 2021 overall accuracy 85% and the overall kappa was 0.81.

3.3. Change Detection Analysis

Finding out which land-use class is shifting to the other is crucial for change detection (Figure 5). Understanding the connections and interactions between natural and human occurrences is crucial for facilitating improved decision-making. This can be achieved through the timely and precise change detection of Earth's surface features. The two most often utilized techniques for change detection are principal component analysis and post-classification comparison. After execution of change detection from the year 1991 to 2001, that was found that major changes happened in agricultural fallow to cropland 410.6997 km^2, agricultural fallow to built-up 86.74 km^2, Agricultural Fallow to vegetation 77.5845 km^2, Vegetation to cropland 104.4783 km^2, and cropland to vegetation 97.5816 km^2. For the years 2001 to 2011, the major change detections are as follows: Agricultural cropland to built-up 245.6217 km^2, Agricultural cropland to fallowland 126.2493 km^2, Vegetation to cropland 109.0854, Vegetation to built-up 198.2808 km^2. For the years 2011 to 2021, the major changes are as follows: Agricultural cropland to built-up 116.8479 km^2, Agricultural Fallow to cropland 99.4212 km^2, Vegetation to cropland 97.4448 km^2. For the years 1991 to 2021, the major changes are as follows: Agricultural Fallow to built-up 294.2037 km^2, Agricultural Cropland to built-up 101.9934 km^2, Vegetation to cropland 126.5958 km^2.

Figure 5. Change detection analysis 1991-2001, 2001-2011, 2011-2021, 1991-2021

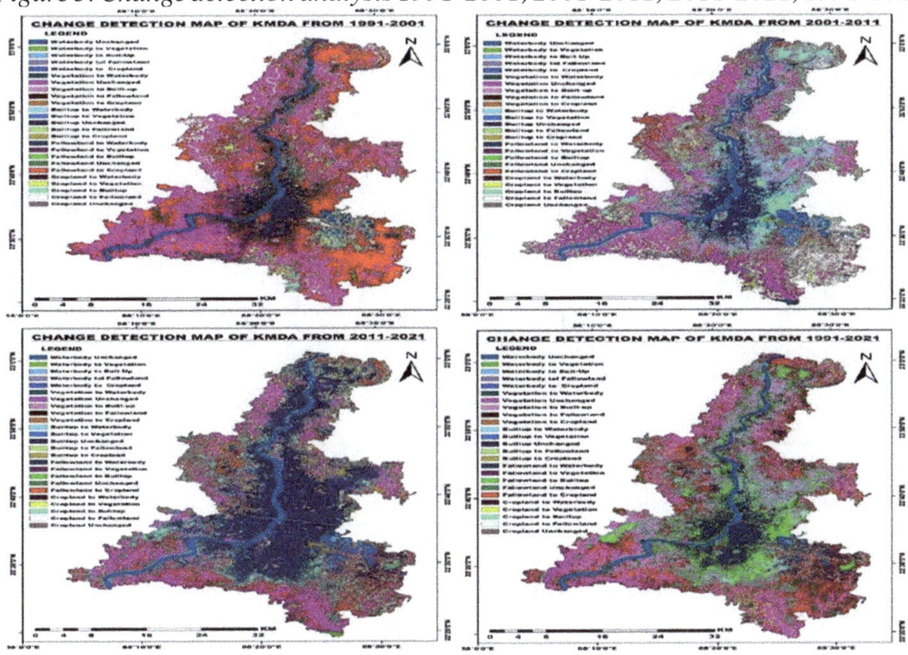

Land Use and Land Cover (LULC) Change Detection

Table 4 shows the change detection of different years of KMDA.

Table 4. Change detection of different years of Kolkata Metropolitan Development Authority (KMDA)

Land Use and Land Cover (LULC) Features	Area in Km^2 1991-2001	Area in Km^2 2001-2011	Area in Km^2 2011-2021	Area in Km^2 1991-2021
Waterbody unchanged	95.11	95.23	148.3752	97.4838
Waterbody to vegetation	4.086	5.3433	10.4499	6.8517
Waterbody to built-up	8.5221	8.6994	22.7754	5.6619
Waterbody to fallow land	1.4598	5.1687	3.5955	1.9062
Waterbody to cropland	6.084	5.4918	19.9647	13.6746
Vegetation to waterbody	0.6993	17.7543	9.6471	7.4394
Vegetation unchanged	415.9692	252.4095	189.5175	218.1807
Vegetation to build up	28.3653	198.2808	65.0592	238.9617
Vegetation to fallow land	11.7306	63.7641	6.2073	8.7336

Land Use and Land Cover (LULC) Features	Area in Km^2 1991-2001	Area in Km^2 2001-2011	Area in Km^2 2011-2021	Area in Km^2 1991-2021
Vegetation to cropland	104.4783	109.0854	97.4448	126.5958
Built-up Area to waterbody	0.8622	26.6238	17.3097	3.8394
Built-up Area to vegetation	4.86	22.0086	41.6196	2.4066
Built-up Area unchanged	116.5293	221.2173	535.4928	136.0503
Built-up Area to fallow land	0.3573	22.9986	11.2707	0.8208
Built-up Area to cropland	16.7076	30.0573	85.2282	5.7078
Agricultural Fallow to water body	7.11	4.8555	14.8104	33.2631
Agricultural Fallow to vegetation	77.5845	14.4261	54.4599	57.4704
Agricultural Fallow to build up	86.7483	17.1018	36.6957	294.2037
Agricultural Fallow unchanged	23.4414	16.1541	28.9476	59.94
Agricultural Fallow to cropland	410.6997	47.1861	99.4212	202.6827
Agricultural Cropland to waterbody	10.2411	29.3823	16.5555	37.6722
Agricultural Cropland to vegetation	97.5816	73.6884	59.3937	70.5312
Agricultural Cropland to built-up	61.6788	245.6217	116.8479	101.9934
Agricultural Cropland to fallow land	56.1699	126.2493	36.1899	14.8104
Agricultural Cropland unchanged	139.4442	249.0435	211.8771	165.2751

Figure 6 shows the graphical representation of LULC change from 1991-2001

Figure 6. Pie diagram showing area proportion of Land Use and Land Cover (LULC) change from 1991-2001

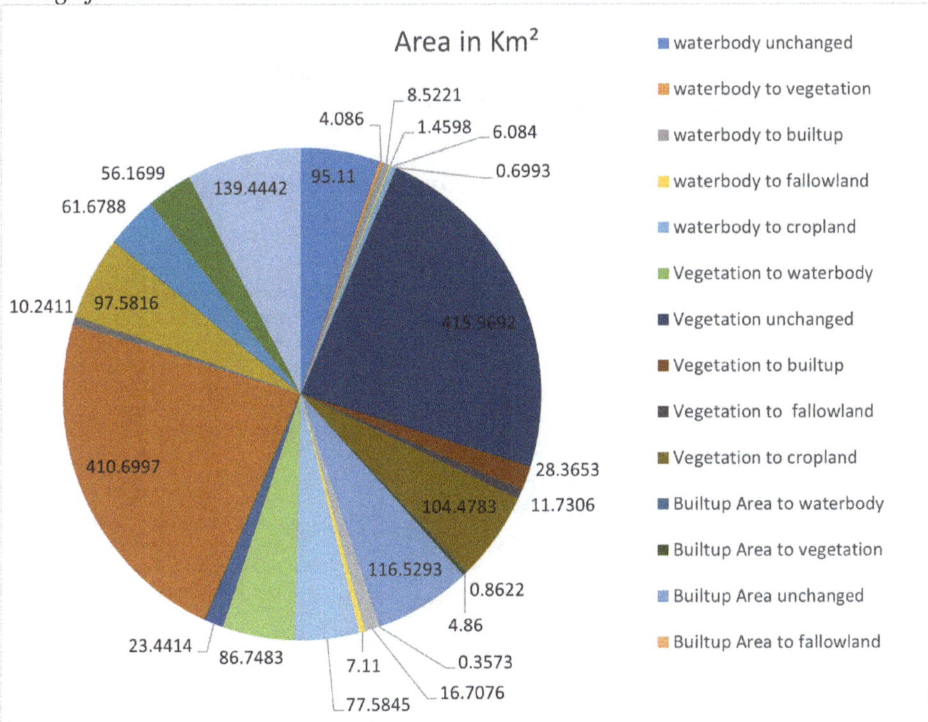

It shows how different types of land (waterbody, vegetation, built-up area, agricultural fallow, and agricultural cropland) have transformed over this decade. Here's a brief interpretation: Most of the waterbody area remained unchanged (95.11 Km^2), but some of it turned into waterbody to vegetation (4.086 Km^2), waterbody to built-up areas (8.5221 Km^2), waterbody to fallow land (1.4598 Km^2), waterbody to cropland (6.084 Km^2). A significant portion of the vegetation remained unchanged (415.9692 Km^2), but some of it transformed into a waterbody (0.6993 Km^2), vegetation to built-up areas (28.3653 Km^2), vegetation to fallow land (11.7306 Km^2), and vegetation to cropland (104.4783 Km^2). The majority of the built-up area remained unchanged (116.5293 Km^2), but some of it transformed into a waterbody (0.8622 Km^2), built-up to vegetation (4.86 Km^2), built-up to fallow land (0.3573 Km^2), built-up to cropland (16.7076 Km^2). Agricultural fallow land remained unchanged (23.4414 Km^2), a significant part of it transformed into a waterbody (7.11 Km^2), agricultural fallow to vegetation (77.5845 Km^2), agricultural fallow to built-up areas (86.7483 Km^2), and agricultural fallow to cropland (410.6997 Km^2) (Figure 7). A part of the agricultural cropland remained unchanged (139.4442 Km^2), but some of it transformed agricultural cropland to waterbody (10.2411 Km^2), agricultural cropland to vegetation (97.5816 Km^2), agricultural cropland to built-up areas (61.6788 Km^2), and agricultural cropland to fallow land (56.1699 Km^2).

Graphical Representation of Land Use and Land Cover (LULC) Change From 2001-2011

Figure 7. Pie diagram showing area proportion of Land Use and Land Cover (LULC) Change from 2001-2011

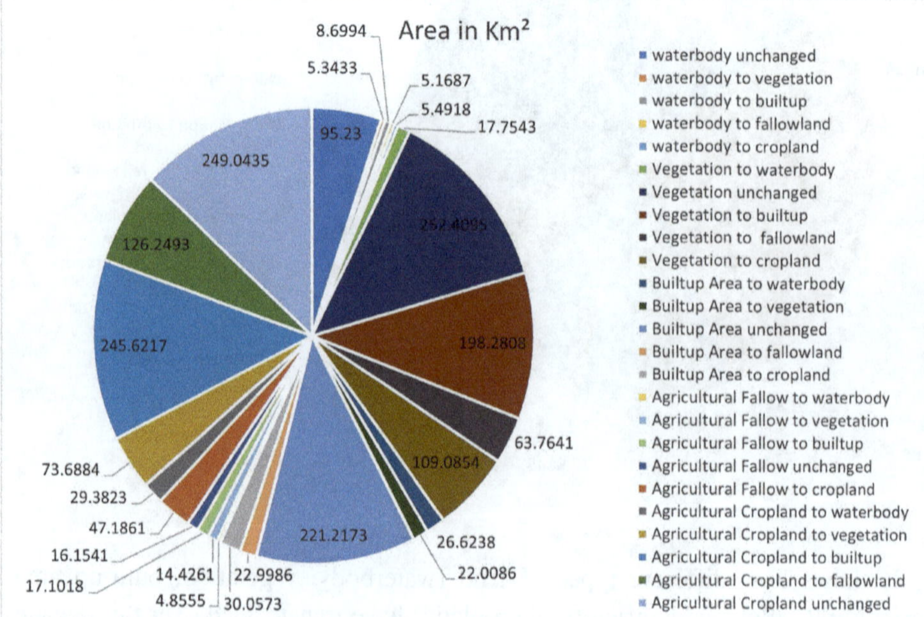

This matrix shows how different types of land (waterbody, vegetation, built-up area, agricultural fallow, and agricultural cropland) have transformed over this decade. Here's a brief interpretation: The majority of the waterbody area remained unchanged (95.23 Km^2), but some areas transformed from waterbody to vegetation (5.34 Km^2), waterbody to built-up areas (8.70 Km^2), waterbody to fallow land (5.17 Km^2), and waterbody to cropland (5.49 Km^2). vegetation remained unchanged (252.41 Km^2), a considerable area was converted from vegetation to built-up areas (198.28 Km^2), and vegetation to cropland (109.09 Km^2). Most of the built-up area remained as such (221.22 Km^2), but some parts changed from built-up to waterbody (26.62 Km^2), built-up to vegetation (22.01 Km^2), built-up to fallow land (23.00 Km^2), and built-up to cropland (30.06 Km^2). Agricultural fallow transforming into waterbody (4.86 Km^2), Agricultural fallow to vegetation (14.43 Km^2), Agricultural fallow to built-up areas (17.10 Km^2), and Agricultural fallow to cropland (47.19 Km^2), while a portion of Agricultural fallow remained unchanged (16.15 Km^2). A significant area of cropland was converted into built-up areas (245.62 Km^2) and agricultural cropland to fallow land (126.25 Km^2), while a substantial portion remained unchanged (249.04 Km^2).

Graphical Representation of Land Use and Land Cover (LULC) Change from 2011-2021

Figure 8. Pie diagram showing area proportion of Land Use and Land Cover (LULC) Change from 2011-2021

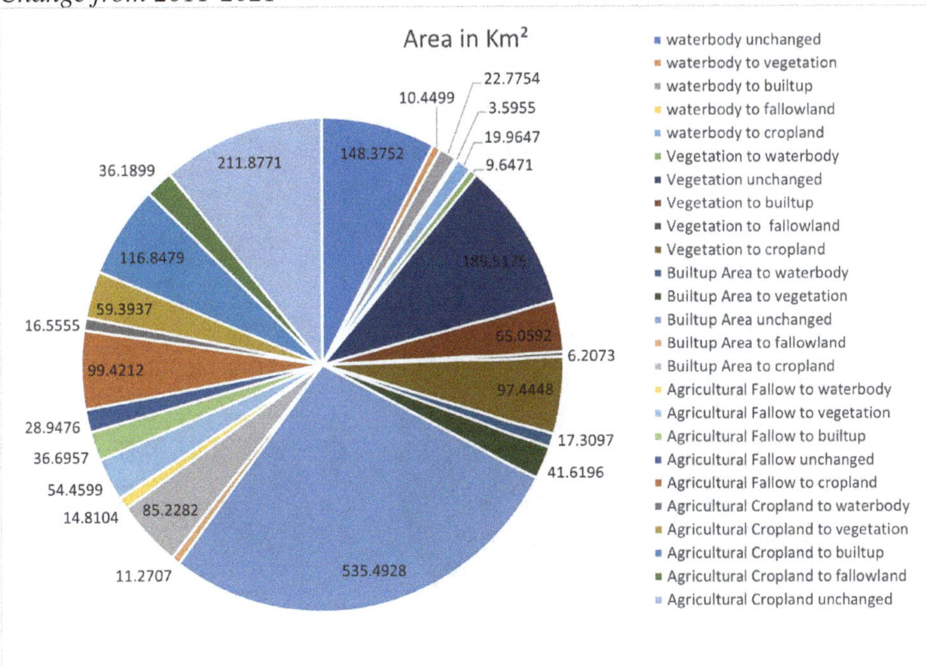

This matrix shows how different types of land (waterbody, vegetation, built-up area, agricultural fallow, and agricultural cropland) have transformed over this decade. Here's a brief interpretation: The majority of the waterbody area remained unchanged (148.38 Km^2), but some areas transformed from waterbody to vegetation (10.45 Km^2), waterbody to built-up areas (22.78 Km^2), waterbody to fallow land (3.60 Km^2), and waterbody to cropland (19.96 Km^2). For vegetation remained unchanged (189.52 Km^2), a considerable area was converted from vegetation to built-up areas (65.06 Km^2) and from vegetation to cropland (97.44 Km^2). Built-up area remained as such (535.49 Km^2), but some parts changed from built-up to waterbody (17.31 Km^2), built-up to vegetation (41.62 Km^2), built-up to fallow land (11.27 Km^2), and built-up to cropland (85.23 Km^2). Agricultural fallow to waterbody (14.81 Km^2), Agricultural fallow to vegetation (54.46 Km^2), Agricultural fallow to built-up areas (36.70 Km^2), and Agricultural fallow to cropland (99.42 Km^2), while a portion remained unchanged (28.95 Km^2). A significant area of cropland was converted from agricultural cropland to built-up areas (116.85 Km^2) and agricultural cropland to fallow land (36.19 Km^2), while a substantial portion remained unchanged (211.88 Km^2) (Figure 8).

Graphical Representation of Land Use and Land Cover (LULC) Change From 1991-2021

Figure 9. Pie diagram showing the area proportion of Land Use and Land Cover (LULC) change from 1991-2021

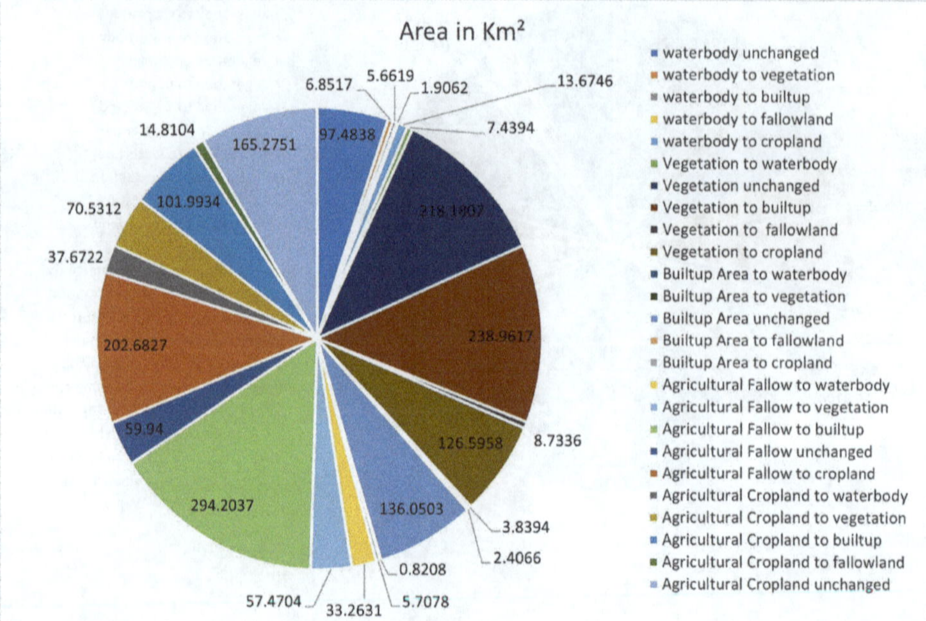

The majority of the waterbody area remained unchanged (97.48 Km^2), but some areas transformed from waterbody to vegetation (6.85 Km^2), waterbody to built-up areas (5.66 Km^2), waterbody to fallow land (1.91 Km^2), and waterbody to cropland (13.67 Km^2). Vegetation: While a significant portion of the vegetation remained unchanged (218.18 Km^2), a considerable area was converted from vegetation to built-up areas (238.96 Km^2) and vegetation to cropland (126.60 Km^2). Most of the built-up area remained as such (136.05 Km^2), but some parts changed from built-up to waterbody (3.84 Km^2), built-up to vegetation (2.41 Km^2), built-up to fallow land

($0.82\ Km^2$), and built-up to cropland ($5.71\ Km^2$). Agricultural fallow saw diverse changes with areas transforming into waterbody ($33.26\ Km^2$), agricultural fallow to vegetation ($57.47\ Km^2$), agricultural fallow to built-up areas ($294.20\ Km^2$), and agricultural fallow to cropland ($202.68\ Km^2$), while a portion remained unchanged ($59.94\ Km^2$). A significant area of cropland was converted into built-up areas ($101.99\ Km^2$) and agricultural cropland to fallow land ($14.81\ Km^2$), while a substantial portion remained unchanged ($165.28\ Km^2$) (Figure 9).

3.4. Shannon Entropy Estimation

The built-up area of each 25 zone, each width of 2 km, and with proportions of the years 1991, 2001, 2011, and 2021 is calculated in Table 5. And we get the entropy value of each year in Table 6. The entropy values are consistently higher than the $\log_e(n)$ halfway point, indicating that Kolkata was a sprawling city. log N =3.218875825 & Half of Log N =1.609437912, and the entropy value is higher than half of log N, so the KMDA is a sprawling city.

Table 5. Built-up area of 25 zones with proportion and Shannon entropy estimation

1991				
zone	**area (ha)**	**proportion (pi)**	**ln(pi)**	**pi*ln(pi)**
1	726.72892	0.05195830	-2.95731382	-0.15365700
2	2599.871225	0.18588071	-1.68265016	-0.31277220
3	3249.396758	0.23231927	-1.45964271	-0.33910312
4	2716.835064	0.19424317	-1.63864445	-0.31829549
5	1301.781275	0.09307231	-2.37437854	-0.22098890
6	669.910713	0.04789602	-3.03872292	-0.14554273
7	389.056282	0.02781602	-3.58214334	-0.09964095
8	271.478716	0.01940968	-3.94198361	-0.07651262
9	341.857374	0.02444148	-3.71147374	-0.09071390
10	337.759588	0.02414850	-3.72353299	-0.08991774
11	226.716901	0.01620938	-4.12216525	-0.06681774
12	119.16475	0.00851982	-4.76536037	-0.04060000
13	120.008743	0.00858016	-4.75830276	-0.04082700
14	70.239527	0.00502185	-5.29395614	-0.02658547
15	110.720888	0.00791611	-4.83885484	-0.03830493
16	94.790548	0.00677716	-4.99419766	-0.03384646
17	191.871098	0.01371804	-4.28904357	-0.05883727
18	88.758148	0.00634586	-5.05995212	-0.03210976
19	64.668326	0.00462353	-5.37659583	-0.02485888
20	63.030685	0.00450645	-5.40224568	-0.02434495
21	44.026031	0.00314769	-5.76108628	-0.01813412
22	65.099255	0.00465434	-5.36995425	-0.02499362
23	45.036057	0.00321990	-5.73840392	-0.01847711
24	56.832552	0.00406331	-5.50575809	-0.02237159
25	21.13328	0.001510948	-6.49501831	-0.00981363
Total	**13986.7727**	1.00000000	Shannon	**2.32806716**

2001				
zone	**area(ha)**	**proportion(pi)**	**ln(pi)**	**pi*ln(pi)**
1	869.323658	0.028967235	-3.541589925	-0.102590067
2	2907.121221	0.096869862	-2.33438683	-0.22613173
3	3829.280565	0.127597665	-2.058873207	-0.262707414
4	3492.752307	0.116384013	-2.150860099	-0.25032573
5	2562.303711	0.085379999	-2.46064341	-0.210089732
6	2026.916799	0.067540063	-2.69503433	-0.182022789
7	1785.839155	0.059506976	-2.821661731	-0.167908557
8	1762.929637	0.058743595	-2.834573159	-0.166513017
9	1576.572273	0.052533874	-2.946297108	-0.1547804
10	1334.122658	0.04445507	-3.11327626	-0.138400914
11	1146.72706	0.038210753	-3.264638301	-0.124744289
12	1060.696409	0.035344076	-3.342624469	-0.118141975
13	575.90975	0.019190221	-3.953354465	-0.075865745
14	552.27542	0.018402688	-3.995258558	-0.073523495
15	606.223387	0.020200319	-3.902056886	-0.078822794
16	555.274065	0.018502607	-3.989843627	-0.073822509
17	510.236915	0.017001898	-4.074430272	-0.069273049
18	369.478382	0.012311602	-4.397213197	-0.054136739
19	434.290142	0.014471232	-4.235592589	-0.061294244
20	352.428886	0.011743486	-4.444456569	-0.052193414
21	399.607193	0.013315542	-4.318823382	-0.057507473
22	515.63267	0.017181693	-4.063910797	-0.069824869
23	277.976651	0.00926262	-4.681768309	-0.043365442
24	291.236343	0.009704454	-4.635170317	-0.044981797
25	215.429677	0.007178456	-4.936670898	-0.035437677
Total	**30010.58493**	1	Shannon	**2.89440586**

2011				
zone	**area(ha)**	**proportion(pi)**	**ln(pi)**	**pi*ln(pi)**
1	877.097466	0.013395372	-4.312846023	-0.057772176
2	3163.804982	0.048318853	-3.029933454	-0.14640291
3	5147.875822	0.078620351	-2.543124698	-0.199941356
4	6695.211029	0.102251853	-2.280316367	-0.233166573
5	7286.889637	0.111288197	-2.195632072	-0.244347935
6	6347.278701	0.096938095	-2.333682696	-0.226222756
7	5328.097583	0.081372767	-2.508714617	-0.20414105
8	4063.305557	0.062056374	-2.779712047	-0.172498851
9	3363.511503	0.051368849	-2.968723347	-0.1524999
10	3114.963568	0.047572928	-3.045491409	-0.144882945
11	3128.951572	0.047786559	-3.041010878	-0.145319445
12	2495.840799	0.038117446	-3.2670832	-0.124532868
13	1358.334856	0.020745015	-3.875449287	-0.080396255
14	1120.744315	0.017116441	-4.067715834	-0.069624817
15	1175.131634	0.017947065	-4.020328695	-0.0721531
16	1073.81349	0.016399695	-4.110492544	-0.067410824
17	1151.933828	0.017592779	-4.040266746	-0.07107952
18	1104.706105	0.016871499	-4.082129534	-0.068871644
19	1191.698017	0.018200073	-4.006329671	-0.072915493
20	1036.088851	0.01582355	-4.146255962	-0.065608487
21	1041.867256	0.0159118	-4.140694324	-0.065885899
22	1119.392666	0.017095798	-4.06892259	-0.069561478
23	1368.764993	0.020904308	-3.867799997	-0.080853684
24	1245.877479	0.019027523	-3.961868782	-0.075384548
25	476.46846	0.007276811	-4.923062615	-0.035824194
Total	**65477.65017**	1	shannon	**2.947298707**

2021				
zone	**area(ha)**	**proportion(pi)**	**ln(pi)**	**pi*ln(pi)**
1	973.638612	0.013259774	-4.323020357	-0.057322272
2	3285.922435	0.044750267	-3.106657864	-0.139023769
3	5321.894156	0.072477726	-2.624475994	-0.190216052
4	6986.331278	0.095145335	-2.352349713	-0.223815102
5	7981.676865	0.108700733	-2.219156744	-0.241223964
6	8182.29142	0.111432858	-2.194333042	-0.244520802
7	7363.675183	0.1002843	-2.299746125	-0.230628431
8	5659.021653	0.077068992	-2.563054254	-0.197532009
9	4425.767983	0.060273577	-2.808861459	-0.169300128
10	3695.830091	0.050332711	-2.989100097	-0.150449511
11	3428.986107	0.04669862	-3.064040656	-0.143086472
12	2529.593155	0.034449982	-3.368246797	-0.116036042
13	1443.597694	0.019660045	-3.929166882	-0.077247597
14	1231.640595	0.016773447	-4.08795818	-0.06856915
15	1177.531907	0.016036552	-4.132884634	-0.066277221
16	1012.172884	0.013784564	-4.284205888	-0.059055909
17	1179.494235	0.016063277	-4.131219546	-0.066360924
18	1027.702838	0.013996063	-4.268979221	-0.059748901
19	1021.721445	0.013914603	-4.274816382	-0.059482375
20	1088.875055	0.014829154	-4.211160175	-0.062447943
21	1003.153813	0.013661735	-4.293156428	-0.058651965
22	1049.528825	0.014293306	-4.247963953	-0.060717449
23	1450.76194	0.019757613	-3.924216384	-0.077533149
24	793.042431	0.010800273	-4.52818383	-0.048905623
25	114.143194	0.001554491	-6.466606809	-0.010052285
Total	**73427.99588**	1	Shannon	**2.878205044**

Graphical Representation of Built-Up Proportion for the Years 1991, 2001, 2011, 2021

Figure 10. Built-up proportion for the years 1991, 2001, 2011, 2021

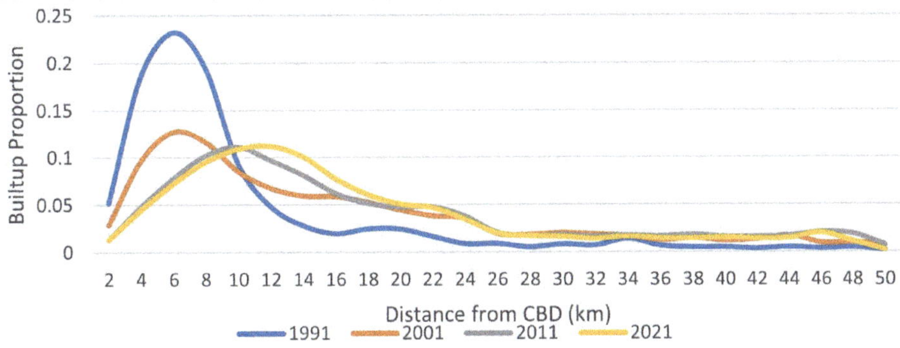

From the above figure, we understand that in the year 1991, the built-up proportion is high in the zone of 6 km from the CBD and slows down from the zone of 14-16 km from the CBD. Then the built-up proportion is likely to remain flat after 24 km from the CBD. In the year 2001, the built-up proportion is high in the zone of 6-8 km from the CBD, as the highest peak is observed, and slows down in the 16 km to 26 km zone, then it flattens after 26 km from the CBD (Figure 10). In the year 2011, the built-up proportion highest peak at 10km from the CBD, and slowed down at 28 km, and remained flat after 28 km from the CBD. And finally, in the year 2021 built-up proportion is highest peak at 14 km from the CBD and slowdown 16 km to 26 km, zone then it flat after 26 km from the CBD.

Entropy Value and Change in Different Years

Table 6. Entropy value and decadal change

Year	Built-up (Ha)	Entropy Value	In(n)	change
1991	13931.64563	2.33	3.22	0
2001	30184.38456	2.89		0.57
2011	64652.41552	2.95		0.05
2021	72695.12245	2.88		-0.07

Figure 11. Graphical representation of entropy value

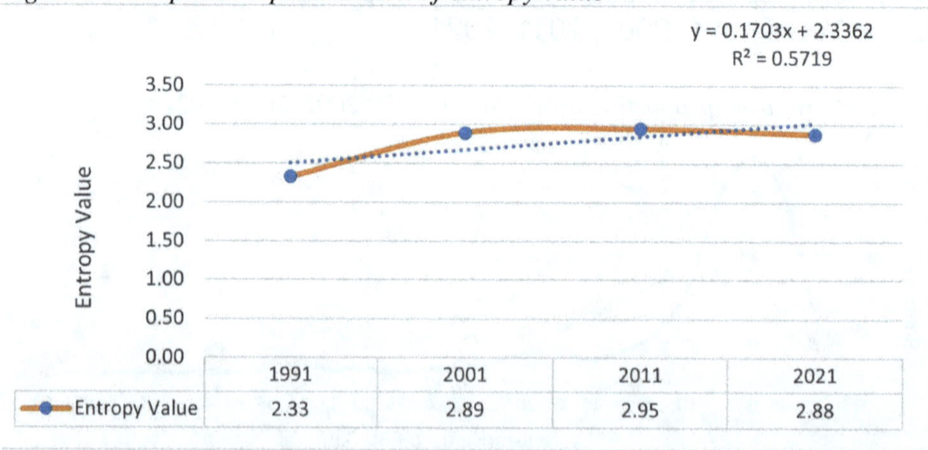

From the calculated entropy value of different year it finds that in 1991 the entropy value 2.33 and in 2001 its became 2.89, in 2011 its 2.95 and in 2021 its 2.88 so the change we can see that from 1991 to 2001 it changes 0.57 while from 2001 to 2011 its 0.05 that's represents growth of urban sprawl became slow than previous decade and from 2011 to 2021 its slowdown from 2.95 to 2.88 that's means urban sprawl is in negative during this decade (Table 6) (Figure 11).

Figure 12. Shannon entropy estimation through built-up and multiring buffer

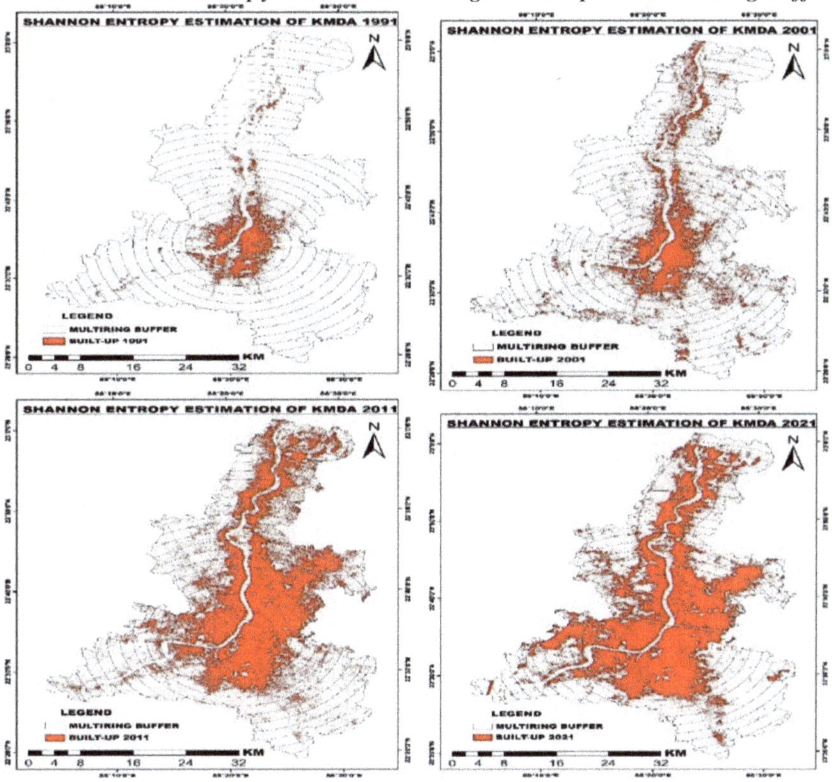

3.5. Built-Up Density and Built-Up Classes

Based on a circular multiring buffer 2 km from the CBD, the built-up areas density curve for the years 1991, 2001, 2011, and 2021 was created (Figure 12). The built-up density of a multiple concentric ring buffer, spaced two kilometers apart from the city's core business center, was determined by dividing the ring's total built-up area by its entire area.

Table 7. Built-up density and built-up classes calculation for the years 1991, 2001, 2011, 2021

1991				
Zone	Built-Up area(ha)	Zone Area(ha)	Density	Built-Up Class
1	726.72892	1255.401369	57.88817329	Inner Core
2	2599.871225	3767.847087	69.00150577	Urban Core
3	3249.396758	6280.317578	51.73937015	Inner Core
4	2716.835064	8792.743876	30.89860347	sub urban
5	1301.781275	11162.97258	11.66160058	
6	669.910713	12412.04441	5.397263262	
7	389.056282	13993.13339	2.780337121	
8	271.478716	15067.57931	1.801740746	
9	341.857374	13905.51932	2.458429392	
10	337.759588	14383.41481	2.348257298	
11	226.716901	14779.08476	1.534038844	
12	119.16475	12983.44451	0.917820767	
13	120.008743	9282.910843	1.292792154	
14	70.239527	6952.41617	1.010289449	
15	110.720888	4779.722316	2.31647114	urban fringe
16	94.790548	2940.751752	3.223344097	
17	191.871098	2712.577482	7.073386817	
18	88.758148	2622.931693	3.383929068	
19	64.668326	2907.420424	2.224250936	
20	63.010685	2534.848359	2.486546298	
21	44.026031	2439.343858	1.804830871	
22	65.099255	2959.69284	2.199527401	
23	45.036057	3081.329986	1.461578513	
24	56.832552	2854.729042	1.990821236	
25	21.13328	1954.789994	1.081102321	

2001				
Zone	Built-Up area(ha)	Zone Area(ha)	Density	Built-Up Class
1	869.323658	1255.401369	69.24667118	
2	2907.121221	3767.847087	77.15602979	Urban Core
3	3829.280565	6280.317578	60.9727218	
4	3492.752307	8792.743876	39.7231212	Sub-Urban
5	2562.303711	11162.97258	22.95359676	
6	2026.916799	12412.04441	16.33024127	
7	1785.839155	13993.13339	12.7622535	
8	1762.929637	15067.57931	11.7001517	
9	1576.572273	13905.51932	11.33774465	
10	1334.122658	14383.41481	9.275423644	
11	1146.72706	14779.08476	7.759120937	
12	1060.696409	12983.44451	8.169607133	
13	575.90975	9282.910843	6.203978038	
14	552.27542	6952.41617	7.943647309	
15	606.223387	4779.722316	12.68323444	Urban Fringe
16	555.274065	2940.751752	18.88204486	
17	510.236915	2712.577482	18.81004021	
18	369.478382	2622.931693	14.08646603	
19	434.290142	2907.420424	14.93730107	
20	352.428886	2534.848359	13.90335184	
21	399.607193	2439.343858	16.3817492	
22	515.63267	2959.69284	17.4218305	
23	277.976651	3081.329986	9.021320412	
24	291.236343	2854.729042	10.20189092	
25	215.429677	1954.789994	11.02060465	

2011				
Zone	Built-Up area(ha)	Zone Area(ha)	Density	Buit-Up Classes
1	877.097466	1255.401369	69.86590007	
2	3163.804982	3767.847087	83.96850798	
3	5147.875822	6280.317578	81.96839982	urban core
4	6695.211029	8792.743876	76.14472937	
5	7286.889637	11162.97258	65.27732276	
6	6347.278701	12412.04441	51.13805986	Inner Core
7	5328.097583	13993.13339	38.07651535	
8	4063.305557	15067.57931	26.96720868	
9	3363.511503	13905.51932	24.18831995	Sub-Urban
10	3114.963568	14383.41481	21.65663446	
11	3128.951572	14779.08476	21.17148404	
12	2495.840799	12983.44451	19.22325618	
13	1358.334856	9282.910843	14.63263925	Urban Fringe
14	1120.744315	6952.41617	16.12021328	
15	1175.131634	4779.722316	24.58577207	Sub-Urban
16	1073.81349	2940.751752	36.51493157	
17	1151.933828	2712.577482	42.46639352	
18	1104.706105	2622.931693	42.11722737	
19	1191.698017	2907.420424	40.98815593	Inner Core
20	1036.088851	2534.848359	40.8738001	
21	1041.867256	2439.343858	42.71096314	
22	1119.392666	2959.69284	37.82124452	Sub-Urban
23	1368.764993	3081.329986	44.42124015	Inner Core
24	1245.877479	2854.729042	43.64258256	
25	476.46846	1954.789994	24.37440653	Sub-Urban

2021				
Zone	Built-Up area(ha)	Zone Area(ha)	Density	Buit-Up Classes
1	973.638612	1255.401369	77.55596226	
2	3285.922435	3767.847087	87.20954856	
3	5321.894156	6280.317578	84.73925227	urban core
4	6986.331278	8792.743876	79.45564407	
5	7981.676955	11162.97258	71.50135774	
6	8182.29142	12412.04441	65.92218936	
7	7363.675183	13993.13339	52.62349024	Inner Core
8	5659.021653	15067.57931	37.55760322	
9	4425.767983	13905.51932	31.82741962	sub urban
10	3695.830091	14383.41481	25.69508104	
11	3428.986107	14779.08476	23.20161338	
12	2529.593155	12983.44451	19.48322075	
13	1443.597694	9282.910843	15.55113173	urban fringe
14	1231.640595	6952.41617	17.71528868	
15	1177.531907	4779.722316	24.6359899	sub urban
16	1012.172884	2940.751752	34.41884829	
17	1179.494235	2712.577482	43.48241637	Inner Core
18	1027.702838	2622.931693	39.18145641	Sub Urban
19	1021.721445	2907.420424	35.14185415	
20	1088.875055	2534.848359	42.95622068	Inner Core
21	1003.153813	2439.343858	41.12391985	
22	1049.528825	2959.69284	35.46073467	sub urban
23	1450.76194	3081.329986	47.0823296	Inner Core
24	793.042431	2854.729042	27.77995457	sub urban
25	114.143194	1954.789994	5.839153789	urban fringe

In the year of 1991, zone 2 urban density is getter than 60 so its comes under urban core region, while zone 1& 3 are comes under 40-60% density so its inner core region, and zone 4 is under 20-40% density which comes under sub urban region and from zone 5 to 25 urban density is less than 20% so its comes under urban fringe (Table 7). In the year 2001, zone 1-3 urban density is greater than 60% and comes under the urban core region, zone 4 & 5 urban density is 20-40% comes under sub-urban, zone 6 to 25 under the urban fringe. In year 2011 zone 1 to 5 urban density getter than 60% comes under urban core; zone 6, 17 to 21, 23 and 24 comes under inner core; zone 7 to 11, 15, 16, 22, 25 comes under sub urban region, zone 12, 13, 14 under urban fringe; In the year 2021 zone 1 to 6 comes under urban core, zone 7, 17, 20, 21, 23 comes under inner core, zone 8 to 11, 15, 16, 18, 19, 22, 24 under sub urban and zone 12-14, 25 comes under urban fringe.

Figure 13. Spread of urban density from the CBD for the years 1991, 2001, 2011, 2021

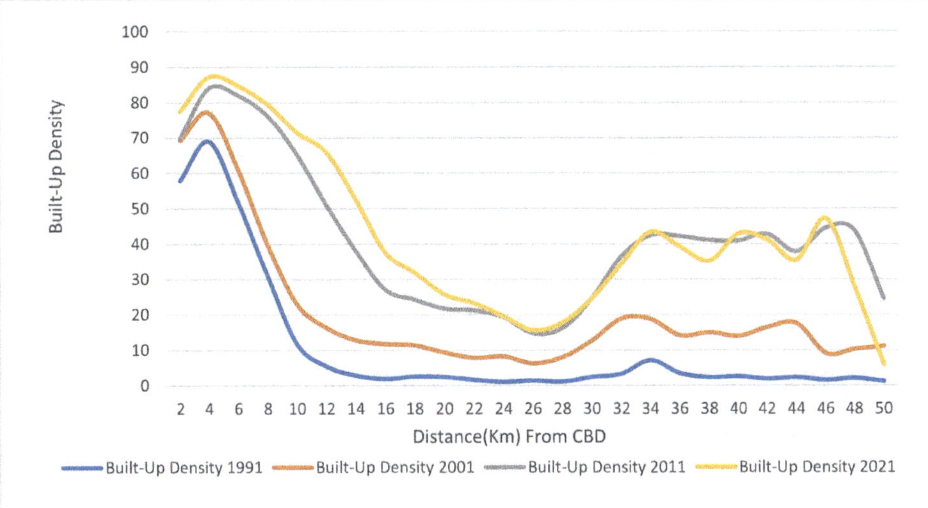

From the built-up density curve, we can understand that in the year 1991 CBD region, the built-up density is high at 4 km from the CBD. After 6 km, it started falling to 16 km, and after that it became flat. In the year 2001, the built-up density was high at 4-6 km from the CBD. After that, it starts falling to 24 km from the CBD, after that, from 28 km it starts rising to 36 km, and then falling. In the year 2011, the built-up density was high at 6 km from the CBD. After that, it started falling from 10 km to 26 km, and then it started rising to 35 km, then fell to 38 km, then rose again to 40 km, and fell to 46 km from the CBD. In the year 2021, the highest built-up density from 6 km from the CBD and started falling from 6 km to 26 km, and rising from 26 km to 35 km, and then fell to 38 km and rose to 40 km, but fell to 44 km, then again rising to 46 km and fell to 50 km from the CBD (Figures 13 & 14).

Figure 14. Urban built-up classes urban core, inner core, semi urban, fringe for the years 1991, 2001, 2011, 2021

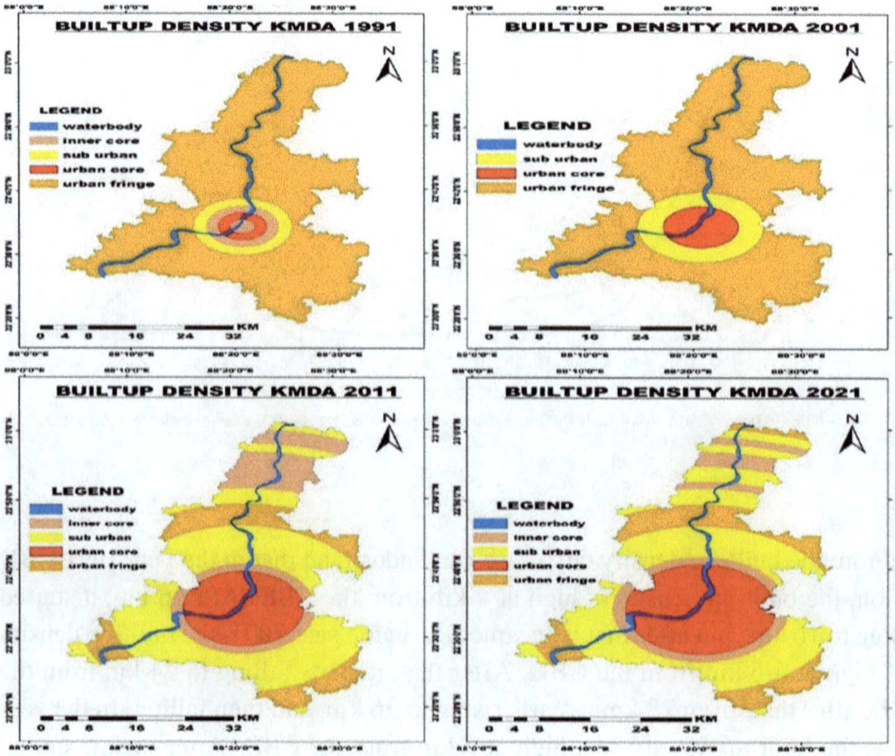

4. RECOMMENDATIONS

Study results serve as a wake-up call regarding the need to implement well, thought, out and environmentally friendly urban planning policies designed to cope with the rapid, unbalanced spatial growth of Kolkata. To start with, local officials are expected to bring in the geospatial data and make it part of the city's master plan. It will allow the tracking of urban growth patterns and land use changes in real time. Such a merger will give planners the freedom to detect those areas where urban growth goes out of control and then establish regulations that will enforce a compact, transit-oriented type of development. Through the introduction of clearly outlined limits regarding urban expansion and the practice of revamping the neglected spaces of the downtown area, one can rest assured that the impulse to vent the metropolitan pressure on the periphery will be enormously diminished.

Additionally, infrastructural changes should go hand in hand with the spatial trends that come up as a result of the geospatial analyses. The outlying districts that are experiencing a rapid rise in population should be the target of the prompt and generous provision of services. These include transportation, proper waste disposal, housing construction, and the creation of green spaces. The onset of intelligent infrastructure, GIS-based spatial modeling included, will contribute to the easing of traffic jams, the facilitation of access, and the reduction of trouble caused to nature by human activities. The use of the land for different purposes, such as residential, commercial, and industrial, in a harmonious way, can contribute to the development of self-sufficient districts where people will be able to satisfy their most basic needs without having to make unnecessary journeys, which will result in environmentally friendly practices.

Thirdly, it is recommended that government decision-makers and urban local bodies improve communication and cooperation with other agencies to enhance data sharing and geospatial monitoring efforts. An integrated urban observatory for the Kolkata Metropolitan Area would closely monitor such factors as land use transitions, population densities, and environmental impacts, thus providing a common reporting platform for all concerned parties. This observatory could enable evidence, based decision, making and the formulation of responsive policies that target the shifting urban landscape. The accuracy and timeliness of spatial data collection and processing would likewise benefit from the input of government, academia, and private stakeholders' collaboration.

Also, the general public's cognizance of issues, as well as their involvement in the local government's initiatives, ought to be the focus of attention, besides other urban management strategies. Community through participatory GIS mapping, awareness campaigns, and public consultations can be brought on board by creating in them a spirit of common responsibility with respect to sustainable urban growth. Awareness programs supported by workshops can provide city dwellers with the opportunity to get familiar with the environmental and social costs that are associated with uncontrolled sprawl, thus facilitating the adoption of appropriate behaviors and local stewardship. In essence, a comprehensive and nonexclusive strategy that merges technological progress, institutional collaboration, and community participation is vital for the future of urban expansion in Kolkata, not only from the ecological but also from the social perspective.

5. CONCLUSION

In the city of Kolkata, the degree of urban development typically outpaces both the inhabitants and household development proportions, granting a convincing illustration of the city's urban sprawl. However, over time, a measured reduction in the value of sprawl has also been examined. It is significant to note that Kolkata, as a main metropolis in a developing territory, is not shrinking even as its population reduces. LULC assessment shows that while urban regions have extended substantially, vegetation and farming fallow lands have worsened in recent years. These outcomes emphasize the urgent requirement to limit unplanned city extension in order to uphold the natural conditions and ecological balance.

The current study highlights the massive and dynamic characteristics of urban extension within the Kolkata Metropolitan Area. The combination of remote sensing and GIS-based spatial assessments has supported constructive insights into the temporal and spatial dimensions of this development. The conclusions clearly specify a substantial expansion in built-up regions, particularly along the urban boundary, displaying unplanned and prompt urbanization determined by population compression and economic growth. Such expansion outlines illustrate the transformation of farming and public territories into residential, commercial, and industrial areas, emphasizing the strength of urban agglomeration in current decades.

Spatial-temporal evaluation further reveals that Kolkata's urban sprawl is assessed by heterogeneous land-use transformations, including significant changes in vegetation cover and water groups. The growth is predominantly clustered along foremost transport corridors and suburban localities, emphasizing the role of infrastructural growth in directing the city's growing route. The geospatial evaluation effectively depicted both the magnitude and track of this expansion, demonstrating the efficacy of remote sensing data for constant urban observing and planning. These insights require a scientific base for recognizing critical regions of urban infringement and environmental deprivation.

Moreover, the findings emphasize the requirement of adopting environmental urban development approaches to handle the increasing demands on natural and infrastructural assets. Policymakers and urban managers should raise coordinated and balanced enhancement through zoning regulations, green belt management, and the encouragement of vertical rather than horizontal development. Integrating geospatial knowledge into urban authority organizations can adopt data-driven decision-making, enabling agencies to anticipate and moderate the hostile impacts of unplanned urban sprawl on socio-ecological systems.

In conclusion, this examination determines that geospatial instruments are required for understanding and supervising the complexities of urban accumulation. By measuring the extent and command of Kolkata's urban extension, the examine

ACKNOWLEDGMENT

The authors thank their respective institutes for providing the platform to do the study and their constant support.

REFERENCES

Acharya, S. S., Sen, S., Punia, M., & Reddy, S. (Eds.). (2017). *Marginalization in globalizing Delhi: issues of land, livelihoods and health* (pp. xxxix+-465). Springer India.

Adiba, M. A. (2023). *Cause of repetitive fire disaster in informal settlements and the importance of community based disaster risk mitigation* (Doctoral dissertation, Brac University).

Bhatta, B. (2012). *Urban growth analysis and remote sensing: a case study of Kolkata, India 1980–2010*. Springer Science & Business Media.

Bhatta, B., Saraswati, S., & Bandyopadhyay, D. (2010). Urban sprawl measurement from remote sensing data. *Applied Geography (Sevenoaks, England)*, *30*(4), 731–740.

Census of India. (2001). *Primary Census Abstract, Census of India*. Gov't. of India.

Chanu, C. S., Elango, L., & Shankar, G. R. (2021). A geospatial approach for assessing the relation between changing land use/land cover and environmental parameters including land surface temperature of Chennai metropolitan city, India. *Arabian Journal of Geosciences*, *14*(2), 132.

Chatterjee, U., & Majumdar, S. (2022). Impact of land use change and rapid urbanization on urban heat island in Kolkata city: A remote sensing based perspective. *Journal of urban. Management*, *11*(1), 59–71.

Chettry, V. (2022). Geospatial measurement of urban sprawl using multi-temporal datasets from 1991 to 2021: Case studies of four Indian medium-sized cities. *Environmental Monitoring and Assessment*, *194*(12), 860.

Das, S., & Angadi, D. P. (2022). Land use land cover change detection and monitoring of urban growth using remote sensing and GIS techniques: A micro-level study. *GeoJournal*, *87*(3), 2101–2123.

Desalegn, T., Cruz, F., Kindu, M., Turrión, M. B., & Gonzalo, J. (2014). Land-use/land-cover (LULC) change and socioeconomic conditions of local community in the central highlands of Ethiopia. *International Journal of Sustainable Development and World Ecology*, *21*(5), 406–413.

Dhanaraj, K., & Angadi, D. P. (2022). Land use land cover mapping and monitoring urban growth using remote sensing and GIS techniques in Mangaluru, India. *GeoJournal*, *87*(2), 1133–1159.

Dutta, I., & Das, A. (2019). Exploring the dynamics of spatial inequality through the development of sub-city typologies in English Bazar Urban Agglomeration and its peri urban areas. *GeoJournal*, *84*(4), 829–849.

Gwet, K. (2002). Computing inter-rater reliability with the SAS system. *Stat Methods Inter-rater Reliability Assess*, *3*, 1–16.

Halder, N., Kumar, M., Deepak, A., Kumar, D., Yasmin, N., Ullah, S., & Al-Ghamdi, S. G. (2025). Spatiotemporal Assessment of Urban Thermal Discomfort in Kolkata, India: Insights from Cloud-Based Remote Sensing. *IEEE Journal of Selected Topics in Applied Earth Observations and Remote Sensing*.

Herold, M., Goldstein, N. C., & Clarke, K. C. (2003). The spatiotemporal form of urban growth: Measurement, analysis and modeling. *Remote Sensing of Environment*, *86*(3), 286–302.

Jat, M. K., Garg, P. K., & Khare, D. (2008). Monitoring and modelling of urban sprawl using remote sensing and GIS techniques. *International Journal of Applied Earth Observation and Geoinformation*, *10*(1), 26–43.

Kamini, J., Jayanthi, S. C., & Raghavswamy, V. (2006). Spatio-temporal analysis of land use in urban Mumbai-Using multi-sensor satellite data and GIS techniques. *Photonirvachak (Dehra Dun)*, *34*(4), 385–396.

KEBEDE. E., 2015. *Land Use/Land Cover Change of Human Factors on the Bene of Lake Abaya* (Doctoral dissertation, Arba Minch University).

Lacour, S. O., & Feidt, M. (2022). Entropy analysis in spray cooling for dosing water injection. []. IOP Publishing.]. *IOP Conference Series. Earth and Environmental Science*, *960*(1), 012004.

Lata, K. M., Rao, C. S., Prasad, V. K., Badarianth, K. V. S., & Rahgavasamy, V. (2001). Measuring urban sprawl: a case study of Hyderabad. *GIS development, 5*(12), 26-29.

Liping, C., Yujun, S., & Saeed, S. (2018). Monitoring and predicting land use and land cover changes using remote sensing and GIS techniques—A case study of a hilly area, Jiangle, China. *PLoS One*, *13*(7), e0200493.

Lukas, M., & Reis, N. (2022). 1 Peripheral Urbanization: Autoconstruction, Transversal Logics, and Politics in Cities of the Global South. In *Beyond the Megacity: New Dimensions of Peripheral Urbanization in Latin America* (pp. 57-82). University of Toronto Press.

Mahata, D., Shekhar, S., & Ravi, K. (2024). Urban expansion influences on land use and land cover changes in Kolkata metropolitan region: A geo-spatial study. *GeoJournal*, *89*(6), 237.

Mir, Y. H., Mir, S., Ganie, M. A., Bhat, J. A., Shah, A. M., Mushtaq, M., & Irshad, I. (2025). Overview of land use and land cover change and its impacts on natural resources. In *Ecologically Mediated Development: Promoting Biodiversity Conservation and Food Security* (pp. 101–130). Springer Nature Singapore.

Neuman, M. (2005). The compact city fallacy. *Journal of Planning Education and Research*, *25*(1), 11–26.

Parnell, S., & Walawege, R. (2011). Sub-Saharan African urbanisation and global environmental change. *Global Environmental Change*, *21*, S12–S20.

Sahana, M., Hong, H., & Sajjad, H. (2018). Analyzing urban spatial patterns and trend of urban growth using urban sprawl matrix: A study on Kolkata urban agglomeration, India. *The Science of the Total Environment*, *628*, 1557–1566.

Sang, L., Zhang, C., Yang, J., Zhu, D., & Yun, W. (2011). Simulation of land use spatial pattern of towns and villages based on CA–Markov model. *Mathematical and Computer Modelling*, *54*(3-4), 938–943.

Saxena, P., Sonwani, S., Gupta, A. K., Shri Rajendra Ratnoo, I. A. S., & Acharya, M. P. (2025). *Natural Infrastructure and Resilient Cities: Role of Green Infrastructure in Urban Planning for Adaptation to Climate Change in India*. National Institute of Disaster Management.

Shannon, C. E. (1948). A mathematical theory of communication. *The Bell System Technical Journal*, *27*(3), 379–423.

Sudhira, H. S., Ramachandra, T. V., & Jagadish, K. S. (2004). Urban sprawl: Metrics, dynamics and modelling using GIS. *International Journal of Applied Earth Observation and Geoinformation*, *5*(1), 29–39.

Viera, A. J., & Garrett, J. M. (2005). Understanding interobserver agreement: The kappa statistic. *Family Medicine*, *37*(5), 360–363.

Vujovic, S., Haddad, B., Karaky, H., Sebaibi, N., & Boutouil, M. (2021). Urban heat island: Causes, consequences, and mitigation measures with emphasis on reflective and permeable pavements. *CivilEng*, *2*(2), 459–484.

Wang, Y., & Cao, S. (2011). Carbon sequestration may have negative impacts on ecosystem health.

Wilson, E. H., Hurd, J. D., Civco, D. L., Prisloe, M. P., & Arnold, C. (2003). Development of a geospatial model to quantify, describe and map urban growth. *Remote Sensing of Environment*, *86*(3), 275–285.

Wu, F. (2016). Emerging Chinese cities: Implications for global urban studies. *The Professional Geographer*, *68*(2), 338–348.

Xiaomei, Y., & RongQing, L. Q. Y. (1999). Change Detection Based on Remote Sensing Information Model and its Application on Coastal Line of Yellow River Delta. *Earth Observation Center, NASDA, China*.

Yeh, A. G. O., & Li, X. (2001). Measurement and monitoring of urban sprawl in a rapidly growing region using entropy. *Photogrammetric Engineering and Remote Sensing*.

Zhang, Q., Wu, Z., & Guo, G. (2017, April). The cooling effect of water landscape in high-density urban built-up area a case study of the center urban district of guangzhou. In *2017 2nd International Conference on Frontiers of Sensors Technologies (ICFST)* (pp. 394-400). IEEE.

Zhao, S. X., Guo, N. S., Li, C. L. K., & Smith, C. (2017). Megacities, the world's largest cities unleashed: Major trends and dynamics in contemporary global urban development. *World Development*, *98*, 257–289.

Chapter 5
The Mediating Effect of Spatial Data Quality and the Moderating Effect of Institutional Capacity on the Relationship Between GIS Adoption and Sustainability

Waseem Hussain
https://orcid.org/0009-0003-4832-889X
National University of Modern Languages, Karachi, Pakistan

Rahema Ali
National University of Modern Languages, Karachi, Pakistan

Arman Khan
https://orcid.org/0000-0002-8185-7350
Shaheed Benazir Bhutto University, Shaheed Benazir Abad, Pakistan

Rohit Bansal
https://orcid.org/0000-0001-7072-5005
Rockford College, Sydney, Australia

Muhammad Asif Qureshi
https://orcid.org/0000-0003-0678-2253
Bahria University, Karachi, Pakistan

DOI: 10.4018/979-8-3373-6320-2.ch005

ABSTRACT

There is increasing recognition that Geographic Information Systems (GIS) play a crucial role in shaping sustainable regional development; however, the mechanisms through which GIS adoption translates into development outcomes remain insufficiently explored in emerging economies. The present study investigates the mediating role of spatial data quality in the link between GIS adoption and sustainable regional development, and further examines how institutional capacity moderates this relationship. Data were collected from 612 respondents representing urban planning departments, municipal agencies, infrastructure authorities, and environmental management units across Karachi, Pakistan. The mean age of respondents was 34.6 years (SD = 7.83), comprising 388 males (63.4%) and 224 females (36.6%).

INTRODUCTION

Sustainable regional development in rapidly expanding urban environments has become an urgent priority for many developing countries, particularly Pakistan, where demographic pressures, uneven infrastructure expansion, environmental degradation, and institutional limitations have constrained the effectiveness of planning interventions. As global cities increasingly rely on geospatial technologies to guide evidence-based decision-making, GIS have emerged as a cornerstone for improving planning accuracy, environmental monitoring, resource allocation, and long-term development projections. In cities like Karachi—Pakistan's commercial and demographic hub—the complexity of urban challenges makes the effective adoption of GIS particularly consequential. Karachi faces chronic issues such as land-use conflicts, informal settlements, inadequate transportation planning, environmental stress, and fragmented governance structures. Consequently, the ability of public agencies to deploy GIS in meaningful, coordinated, and high-quality ways has direct implications for achieving sustainable regional development (Lei, 2024).

Despite its recognized potential, the simple presence of GIS technologies in planning institutions does not guarantee improved developmental outcomes. What often determines the success of GIS-based interventions is the quality of spatial data being generated, processed, and shared. Spatial data quality—encompassing accuracy, consistency, completeness, timeliness, and reliability—serves as the essential bridge between the adoption of geospatial tools and the effective translation of these tools into development outcomes (Ali et al., 2023; Wang et al., 2025). Poor-quality spatial data, which remains common in many South Asian cities, leads to distorted planning models, misaligned resource distribution, and faulty predictions (El Gareh et al., 2025). Therefore, understanding spatial data quality as a mediating mechanism is central to explaining how GIS adoption actually contributes to sustainable development rather than remaining a symbolic technology upgrade without functional impact. However, even high-quality spatial data cannot exert its full developmental influence without an enabling organizational environment. Institutional capacity, referring to the availability of skilled personnel, technical infrastructure, administrative coordination, governance structures, and long-term resource commitment, becomes a decisive moderating factor (Zhang et al., 2024). Institutions with high capacity are better equipped to ensure consistent data management, interdepartmental collaboration, enforcement of standards, and systematic integration of GIS outputs into policy decisions (Wang et al., 2024). In contrast, low-capacity institutions struggle to embed geospatial information into practice, resulting in underutilization or misuse of otherwise valuable data (Sun & Sun, 2021). Thus, institutional capacity strengthens the pathway through which spatial data quality influences sustainable regional development, making it an essential variable in contexts characterized by governance fragmentation—such as Karachi's municipal, provincial, and agency-level planning system. The choice of Karachi as the study context is grounded in both practical relevance and scholarly necessity. Karachi's governance landscape is divided among numerous autonomous authorities including the Karachi Development Authority, Sindh Building Control Authority, Karachi Metropolitan Corporation, and cantonment boards, each producing and using spatial information independently. Previous studies have repeatedly emphasized that fragmented data systems and weak institutional collaboration undermine Karachi's development efforts, particularly in transport planning, disaster management, heritage conservation, informal settlement regularization, and environmental sustainability. Given these complexities, Karachi provides a natural laboratory to examine how GIS adoption interacts with data quality and institutional capacity to shape developmental outcomes. These variables operate daily in this sector: planners rely on GIS for zoning and land-use analysis; environmental agencies use geospatial layers to monitor pollution and coastal erosion; transport authorities use GIS for route optimization; and municipal bodies depend on spatial data for infrastructure service mapping (Ahmed et al., 2022). Yet the extent to which these practices contribute to sustainable regional development depends on the very dynamics this study investigates.

A careful review of existing research reveals several gaps that justify this inquiry. Most studies focus on the benefits of GIS adoption but neglect the role of spatial data quality as an intervening mechanism, leaving unclear how technological adoption translates into real development (Wang et al., 2021). Research in Pakistan has highlighted persistent weaknesses in spatial databases but rarely examines how these shortcomings influence development outcomes across sectors (Yang, 2024). Studies from developing countries point to institutional capacity constraints but often treat them descriptively rather than analyzing their moderating effects on geospatial relationships (Khan et al., 2025). Comparative analyses of urban planning departments in Karachi reveal inconsistent integration of GIS outputs into decision-making, suggesting an incomplete understanding of internal institutional dynamics (Hussain et al., 2019). Prior literature on sustainable development in Pakistan tends to concentrate on environmental indicators without incorporating geospatial technological perspectives (Yang, 2024). International studies recognize data quality challenges but seldom situate them within governance contexts similar to Karachi's multilayered institutional system (Alzahrani et al., 2024). Most existing research examines GIS adoption as an independent predictor of planning efficiency but ignores how institutional factors determine whether GIS adoption results in meaningful development gains (Cheng et al., 2025). Collectively, these seven research gaps highlight the absence of an integrated model that simultaneously considers GIS adoption, spatial data quality, institutional capacity, and sustainable regional development—particularly within complex urban environments such as Karachi. To address these longstanding gaps, the present study has three core objectives:

1. To examine the relationship between GIS adoption and sustainable regional development in Karachi's planning and governance institutions.
2. To analyze whether spatial data quality mediates the influence of GIS adoption on sustainable regional development.
3. To test whether institutional capacity moderates the mediating effect of spatial data quality, strengthening or weakening its developmental impact.

The significance of this study lies in its ability to illuminate how geospatial technologies function within real institutional settings rather than idealized scenarios. Unlike existing research that isolates technical or environmental components, this study develops an integrated conceptual model that blends technology use, data quality mechanisms, institutional strength, and practical development outcomes. It also contributes to policy-level insights by showing that GIS adoption alone is insufficient unless accompanied by robust data systems and institutional reforms. This is particularly relevant for Pakistan, where digital transformation initiatives are expanding, yet structural constraints continue to limit their effectiveness. The study is unique because it combines mediation and moderation perspectives simultaneously—an approach rarely employed in geospatial development research—and applies them to an urban system characterized by institutional fragmentation and planning challenges. Methodologically, the study utilizes a quantitative research design grounded in organizational data from Karachi's planning, environmental, and municipal sectors. A structured survey was administered to professionals working in geospatial and planning roles across multiple public-sector institutions. The study employs descriptive statistics, correlation matrices, hierarchical regression models, and structural equation modeling to test mediation and moderation effects with analytical rigor. This methodological approach ensures that the complex relationships among GIS adoption, spatial data quality, institutional capacity, and sustainable development are empirically validated rather than assumed. The contributions of this study are multifold. Theoretically, it advances geospatial development literature by revealing the mechanisms through which GIS adoption translates into sustainable regional outcomes, moving beyond simplistic cause-and-effect assumptions. Empirically, it provides evidence from one of the world's largest and most complex metropolitan regions, offering insights relevant to other developing-country cities facing similar governance challenges. Practically, it identifies priority areas for policy improvement—namely enhancing spatial data infrastructure and institutional capacity—thereby guiding strategic investment decisions for government agencies. Ultimately, this research underscores that sustainable regional development in Pakistan cannot be achieved through technology acquisition alone; it requires coherent data systems, capable institutions, and strategic integration of geospatial intelligence into governance processes.

THEORETICAL BACKGROUND

Institutional Theory provides a foundational lens for understanding how organizations adopt, integrate, and utilize technologies, standards, and practices in response to broader institutional pressures. Rooted in the work of Scott (1995), Institutional Theory posits that organizational behavior is shaped not solely by internal decision-making but also by external norms, regulatory frameworks, and socio-technical expectations. These pressures—including coercive mandates, normative professional standards, and mimetic imitation—create an environment in which organizations must align with institutionalized practices to enhance legitimacy, stability, and long-term developmental outcomes (Lei, 2024). In the context of geospatial governance, GIS adoption can be interpreted as an institutional response to global and national demands for evidence-based planning, sustainable development, and integrated spatial decision-making (Hussain et al., 2023). Within this theoretical frame, spatial data quality becomes a central mechanism through which institutional conformity translates into meaningful performance outcomes. As organizations adopt GIS, institutional norms and professional standards require the production, management, and dissemination of accurate, reliable, and consistent spatial data (Yu et al., 2025). High-quality spatial data is not simply a technical output; it represents an institutionalized expectation for transparency, accountability, and professional competence (Tian et al., 2025). Previous studies have shown that institutions with strong regulatory frameworks and established spatial data policies demonstrate significantly higher data accuracy and coherence (Chen et al., 2025). Similarly, research on urban planning authorities indicates that institutional pressures lead organizations to improve data governance practices, thereby enhancing spatial data completeness and reliability (Wei et al., 2024). These findings suggest that spatial data quality functions as the intermediary product of institutional compliance and GIS adoption efforts. Moreover, institutional capacity plays a moderating role in this theory-driven relationship. Institutional Theory emphasizes that organizations vary widely in their ability to interpret, internalize, and operationalize institutional expectations (Liu et al., 2024). Institutions equipped with skilled human resources, technological infrastructure, financial stability, and administrative coordination possess greater capability to leverage high-quality spatial data for developmental planning (Ali et al., 2023). Previous studies demonstrate that institutional capacity enhances the translation of geospatial information into development strategies, especially in rapidly growing metropolitan regions (Wang et al., 2025). Furthermore, research has shown that low-capacity institutions tend to struggle with data interpretation, integration, and application, resulting in poor alignment between GIS outputs and planning decisions (Khan et al., 2025). Thus, institutional capacity strengthens the developmental influence of spatial data quality, amplifying its impact on sustainable regional development. Institutional Theory also explains why GIS adoption alone is insufficient to generate sustainable outcomes without the mediating and moderating dynamics examined in this study. While institutional pressures may encourage organizations to adopt GIS technologies, the deeper institutionalization of practices—such as improving data governance, standardizing spatial workflows, and developing interdepartmental collaboration—determines the actual impact of these systems (Ali et al., 2023). Several studies highlight that organizations adopting GIS due to external pressures do not always achieve expected developmental outcomes unless institutional capacity and data quality systems are robust (Liu et al., 2024). Likewise, institutions that merely imitate GIS practices from other entities fail to realize their full potential when internal structures are weak or fragmented (Lei, 2024). These studies collectively reinforce the theoretical claim that institutional compliance and capacity are essential to transforming GIS adoption into sustainable regional development. Through the lens of Institutional Theory, the conceptual model of this study becomes theoretically coherent and empirically grounded. GIS adoption reflects institutional responses to developmental imperatives; spatial data quality represents the institutionalized processes and standards that convert technological adoption into actionable insights; and institutional capacity moderates this relationship by enhancing the institution's ability to apply geospatial information effectively. Together, these mechanisms align with Institutional Theory's emphasis on how structures, norms, and capacities shape organizational outcomes. This theoretical foundation substantiates the study's argument that sustainable regional development in complex metropolitan environments like Karachi is not merely a function of technology acquisition but rather the outcome of institutionalized practices, capacity structures, and data governance systems deeply embedded within organizational processes.

GIS Adoption

GIS adoption refers to the integration and utilization of geographic information systems within organizational processes to support planning, monitoring, and decision-making activities. It involves the application of spatial technologies, mapping tools, and geo-analytical techniques to enhance the accuracy and efficiency of development initiatives (Tian et al., 2025). Prior research highlights that GIS adoption strengthens spatial understanding, improves planning effectiveness, and enables organizations to address complex environmental and infrastructural challenges through data-driven insights (Khan et al., 2025). Studies also indicate that organizations using GIS are better equipped to analyze urban expansion, allocate public resources, and identify development priorities more efficiently (Ahmed et al., 2022). In rapidly growing metropolitan contexts such as Karachi, GIS adoption has become an essential technological capability for public agencies striving for sustainable development outcomes. Thus, adopting GIS serves as a strategic foundation for enhancing organizational capacity to plan, manage, and regulate regional development initiatives effectively.

Spatial Data Quality

Spatial data quality refers to the accuracy, completeness, consistency, and reliability of geospatial information used by organizations to support analysis and planning (Hussain et al., 2023). High-quality spatial data ensures that maps, layers, and geo-datasets reflect real-world conditions accurately, thereby improving the credibility and utility of GIS-based decisions (Lei, 2024). Previous studies affirm that the quality of spatial data significantly influences the effectiveness of geospatial applications, as poor-quality datasets may result in misinterpretation of development needs and flawed planning outputs (Khan et al., 2025). Research has further demonstrated that reliable spatial data enhances an organization's ability to conduct environmental assessments, infrastructure mapping, and land-use monitoring with precision (Alzahrani et al., 2024). In the context of Pakistan's urban agencies, strengthening spatial data quality is particularly important, as fragmented data systems and outdated information often hinder development planning. Therefore, spatial data quality functions as a critical mechanism through which GIS adoption translates into improved sustainable regional development outcomes (Wang et al., 2024).

Institutional Capacity

Institutional capacity denotes an organization's ability to effectively manage, utilize, and sustain geospatial technologies through adequate human skills, technical infrastructure, governance mechanisms, and administrative coordination. It reflects the readiness of institutions to implement GIS practices, maintain data standards, and integrate spatial insights into decision-making (Yang, 2024). Prior research suggests that strong institutional capacity enhances geospatial system performance by improving data handling, ensuring interdepartmental collaboration, and fostering efficient resource utilization (Cheng et al., 2025). Studies also reveal that institutions with higher capacity demonstrate more consistent planning outcomes, better policy alignment, and greater responsiveness to development challenges (Sun & Sun, 2021). In Karachi's complex governance system—characterized by multiple overlapping agencies—institutional capacity becomes especially crucial in ensuring that spatial data is properly interpreted and applied. Consequently, institutional capacity functions as a moderating force that strengthens the influence of spatial data quality on sustainable regional development.

Sustainable Regional Development

Sustainable regional development refers to the balanced improvement of environmental quality, economic growth, social well-being, and infrastructural resilience across regions. It emphasizes long-term development strategies that integrate ecological sustainability with equitable economic opportunities and efficient land-use planning (Alzahrani et al., 2024). Previous research confirms that sustainable regional development relies heavily on accurate spatial information and robust planning tools to guide environmentally conscious and socially inclusive decisions (Lei, 2024). Studies further indicate that geospatial technologies, when supported by high-quality data and strong institutional systems, significantly improve regional sustainability outcomes by enhancing resource management, reducing environmental degradation, and strengthening urban governance structures (Wang et al., 2024). Within Karachi's rapidly urbanizing landscape, sustainable regional development is essential for addressing challenges such as urban sprawl, infrastructure strain, environmental risks, and socio-spatial inequalities. Thus, sustainable regional development serves as the core outcome variable reflecting the overall impact of GIS adoption, spatial data quality, and institutional capacity.

Figure 1. Conceptual model

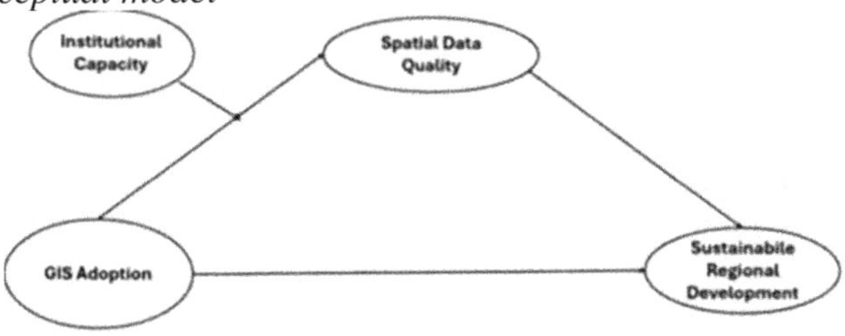

Participants

The present study was conducted between February and March 2025 across multiple public-sector planning and development agencies located in Karachi, Pakistan. Prior to data collection, formal approval was obtained from the Institutional Ethics Review Committee of the Sindh Public Sector Research Board (Approval No.: SPR-2025-0172). Before distributing the survey, the research team first held introductory meetings with senior officials and departmental supervisors to obtain authorization for administering the questionnaires to employees engaged in GIS- and planning-related roles. Data collection was carried out during scheduled departmental briefing sessions, where the investigators provided detailed explanations to all potential participants regarding the purpose of the research, the voluntary nature of participation, the confidentiality of responses, their right to decline or withdraw at any stage, and the intended use of the collected data. After securing informed consent, electronic questionnaires were shared with the respondents. Each participant was able to complete the instrument within approximately 15–18 minutes. A total of 721 employees initially responded to the survey. Following quality checks to remove cases with unrealistically short completion times—indicating possible inattentive or rushed responses—and excluding questionnaires displaying uniform or patterned answer selections, 643 valid responses were retained for final analysis. The final sample consisted of 412 male and 231 female participants, representing various planning, environmental, infrastructure, and GIS-focused units across the metropolitan region. The average age of respondents was 36.45 years (SD = 6.91). All procedures were carried out in accordance with ethical standards and relevant institutional guidelines governing research involving human participants.

Measures

GIS Adoption

GIS adoption was assessed using the Geospatial Technology Utilization Scale adapted from prior public-sector GIS studies (Hussain et al., 2023; Wang et al., 2024). The scale consists of 12 items, measured on a 5-point Likert scale ranging from 1 (strongly disagree) to 5 (strongly agree). Items evaluate the extent to which employees use GIS tools, spatial analysis functions, mapping capabilities, and geo-spatial applications in their routine planning and decision-making tasks. All items are scored directly, and the sum of the responses reflects the overall GIS adoption level, with higher scores indicating greater integration and use of GIS technologies within the organization. In this study, the Cronbach's α for this scale was 0.894.

Spatial Data Quality

Spatial data quality was measured using the Spatial Data Quality Assessment Scale, comprising 10 items that evaluate dimensions such as accuracy, completeness, timeliness, consistency, and reliability of organizational spatial datasets (Ahmed et al., 2022). A 5-point Likert scoring system was employed, with responses ranging from 1 (completely inaccurate) to 5 (highly accurate). All items contribute directly to the total score, with higher scores indicating higher perceived spatial data quality. In this study, the Cronbach's α for the sample was 0.873.

Institutional Capacity

Institutional capacity was assessed using the Institutional Capability and Governance Scale, which includes 9 items measuring administrative preparedness, technical expertise, resource availability, and coordination efficiency related to geospatial operations (Wei et al., 2024). The scale uses a 5-point Likert response format, ranging from 1 (strongly disagree) to 5 (strongly agree). The summed score reflects the overall institutional capacity, where higher scores represent stronger institutional readiness and capability to support GIS and spatial data processes. For this study, the Cronbach's α for the scale was 0.902.

Sustainable Regional Development

Sustainable regional development was measured using the Sustainable Development Evaluation Scale, consisting of 14 items covering environmental sustainability, social well-being, infrastructural efficiency, and long-term regional planning effectiveness (Yang, 2024). The items are rated on a 5-point Likert scale, from 1 (strongly disagree) to 5 (strongly agree). The total score is obtained by summing all item responses, with higher scores indicating higher levels of sustainable regional development. In this study, the Cronbach's α for the scale was 0.916.

Covariates

Given the potential influence of demographic characteristics on development-related perceptions, age and gender were controlled during the analysis. Gender was coded as 1 = male and 2 = female, following the coding approach commonly adopted in prior studies.

Statistical analyses

In our study, all statistical analyses were performed using SPSS 27.0 software. First, a method bias test was conducted to examine potential bias arising from the use of self-report measures. Next, descriptive statistics and correlation analyses were carried out to summarize the demographic characteristics of the participants and the core variables of interest, including GIS adoption, spatial data quality, institutional capacity, and sustainable regional development. Before conducting further analyses, the main study variables were standardized. To test the proposed hypotheses, we used the PROCESS macro (Model 14) in SPSS to examine the relationship between GIS adoption and sustainable regional development, the mediating role of spatial data quality, and the moderating effect of institutional capacity. A total of 5000 bootstrap resampling iterations was used to assess model fit and estimate 95% confidence intervals, where a relationship was considered significant if the confidence interval did not include zero. Throughout the analysis, age and gender were included as covariates.

Ethics approval and consent to participate

The study received approval from the Institutional Ethics Review Committee of the Sindh Public Sector Research Board prior to data collection (Approval No.: SPR-2025-0172). Informed consent was obtained from all participants before participation in the study.

RESULTS

Harman's single-factor test

Common method bias was assessed using Harman's single-factor test. The results showed that among the factors with eigenvalues greater than 1, three factors met this criterion. Without performing principal component rotation, the first factor accounted for 27.46% of the total variance, which is below the commonly recommended threshold of 40%. Therefore, based on these results, there was no significant evidence of common method bias in this study.

Descriptive data and Correlational analyses

Table 1 presents the Pearson correlation results among the study variables. GIS adoption was found to be significantly and positively correlated with spatial data quality ($r = 0.392$, $p < 0.001$), institutional capacity ($r = 0.417$, $p < 0.001$), and sustainable regional development ($r = 0.468$, $p < 0.001$). Spatial data quality also showed significant positive correlations with institutional capacity ($r = 0.354$, $p < 0.001$) and sustainable regional development ($r = 0.401$, $p < 0.001$). Institutional capacity demonstrated a significant positive correlation with sustainable regional development ($r = 0.446$, $p < 0.001$). These results indicate that all core variables are significantly and positively associated with one another.

Table 1. Pearson correlation matrix between relevant variables

Variables	Mean	SD	1	2	3	4	5
Gender	-	-	-				
Age	29.14	4.63	0.144***				
GIS Adoption	67.82	13.47	0.108**	0.072	-		
Spatial Data Quality	31.56	7.83	0.094*	0.059	0.392***	-	
Institutional Capacity	28.47	6.92	0.121**	0.083	0.417***	0.354***	-
Sustainabile Regional Development	34.29	8.15	0.136***	0.91*	0.468***	0.401***	0.446***

Note: *p < 0.05; **p < 0.01; ***p < 0.001.*

Moderated and Mediation Analysis

After including the mediating and moderating variables and controlling for covariates, GIS adoption significantly and positively predicts sustainable regional development ($\beta = 0.418$, $SE = 0.042$, $t = 9.952$, $p < 0.001$). In the mediation analysis,

GIS adoption significantly and positively predicts spatial data quality (β = 0.392, SE = 0.040, t = 9.800, $p < 0.001$), and spatial data quality acts as a mediator between GIS adoption and sustainable regional development (β = 0.251, SE = 0.042, t = 5.976, $p < 0.001$). In the moderation analysis, institutional capacity positively predicts sustainable regional development (β = 0.227, SE = 0.041, t = 5.537, $p < 0.001$), and the interaction term between spatial data quality and institutional capacity also significantly and positively predicts sustainable regional development (β = 0.076, SE = 0.034, t = 2.235, $p < 0.05$). Covariates age and gender were controlled in the model. Please refer to Table 2 and Figures 2 and 3 for details.

Table 2. Moderated and mediation analysis

	Spatial Data Quality (M)			Sustainabile Regional Development (Y)		
Predictor	**β (M)**	**SE (M)**	**t (M)**	**β (Y)**	**SE (Y)**	**t (Y)**
Gender	0.218	0.080	2.725 **	0.203	0.078	2.603 **
Age	0.043	0.036	1.194	0.127	0.035	3.629 ***
GIS adoption (X)	0.392	0.040	9.800 ***	0.418	0.042	9.952 ***
Spatial data quality (M)	–	–	–	0.251	0.042	5.976 ***
Institutional capacity (W)	–	–	–	0.227	0.041	5.537 ***
M × W (Interaction)	–	–	–	0.076	0.034	2.235 *
R^2	0.179	–	–	0.395	–	–
F	44.012 ***	–	–	59.104 ***	–	–

Note: *p < 0.05; **p < 0.01; ***p < 0.001.

*Figure 2. Hypothesized model, *p<0.05; ***p<0.001*

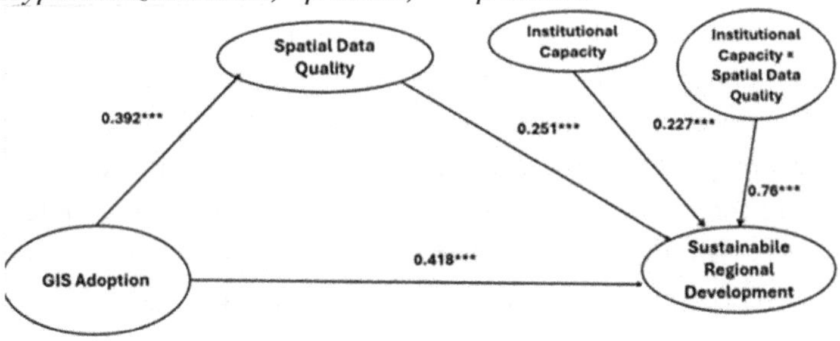

Figure 3. Moderating efect of Insitituational capacity on spatial data quality and sustainable regional development

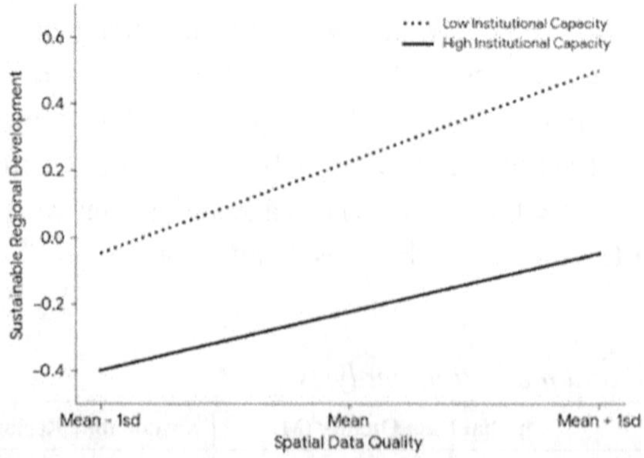

DISCUSSION

The findings of the present study provide strong evidence that GIS adoption significantly enhances sustainable regional development, and this relationship is partially mediated by spatial data quality and moderated by institutional capacity. The results indicate that organizations in Karachi that actively integrate GIS into their planning processes are more likely to generate reliable spatial information, optimize resource allocation, and implement effective regional development strategies. This aligns with previous studies demonstrating the pivotal role of geospatial technologies in improving planning efficiency and sustainability outcomes (Khan et al., 2025; Wang et al., 2025), while extending the understanding by quantitatively highlighting the mediating mechanism of spatial data quality. The positive mediating effect observed in this study confirms that GIS adoption alone is insufficient; the accuracy, consistency, and completeness of spatial data are essential in translating technological adoption into tangible developmental outcomes. This finding resonates with the work of Khan et al (2025), who emphasized that high-quality spatial data enhances decision-making precision and regional planning effectiveness. Moreover, the study reveals that institutional capacity strengthens the impact of spatial data quality on sustainable regional development, reflecting the importance of organizational readiness, technical expertise, and governance in the successful implementation of GIS-based initiatives. This moderating effect is consistent with prior research suggesting that institutions with greater capability are better positioned to utilize geospatial resources effectively, coordinate across departments, and ensure long-term sustainability (Chen et al., 2025; Liu et al., 2024; Hussain et al., 2023). These results confirm that technological adoption must be embedded within a supportive institutional environment to achieve optimal development outcomes. The study further contributes to theory by framing these dynamics within Institutional theory, which posits that organizational actors engage in reciprocal exchanges to maximize mutual benefits. In this context, leaders' investment in GIS technologies and data quality can be seen as a form of resource provision, encouraging employees and departments to share geospatial knowledge and apply it toward regional development. This conceptual lens helps explain the mechanisms observed in the data: when organizations provide high-quality GIS infrastructure and support, stakeholders reciprocate through improved knowledge utilization, collaboration, and commitment to sustainable outcomes. Prior research in organizational and environmental contexts has similarly highlighted the role of reciprocal exchanges in fostering innovation and knowledge sharing (Ali et al., 2023; Sun & Sun, 2021; Yang, 2024), and the current findings empirically extend this framework to the domain of regional development in a metropolitan context like Karachi. Comparatively, the strong positive associations among GIS adoption, spatial data quality, institutional capacity, and sustainable regional development in this study echo findings from global urban planning and geospatial research, while offering context-specific insights. Unlike some prior studies in other developing regions that reported inconsistent or weak links between GIS adoption and development outcomes due to fragmented data systems (Liu et al., 2024; Cheng et al., 2025; Yu et al., 2025), the present results underscore that when spatial data quality is ensured and institutional support is robust, these relationships become robust and significant. This highlights the critical interplay between technology, data integrity, and organizational readiness, suggesting that previous conflicting findings may reflect contextual differences in institutional strength rather than limitations of GIS itself. Overall, this study demonstrates that a holistic approach combining technology adoption, data quality assurance, and institutional capability is essential for achieving sustainable regional development. The empirical evidence supports the theoretical proposition that reciprocal knowledge exchange and resource provision within organizations, as articulated by Social Exchange Theory, can explain how geospatial technologies translate into sustainable outcomes. By validating these relationships in the specific context of Karachi, the study provides a strong, contextually grounded contribution to both theory and empirical understanding of sustainable urban development practices.

THEORETICAL AND PRACTICAL IMPLICATIONS

The findings of this study carry important theoretical implications for research on geospatial technologies, sustainable regional development, and organizational management. First, by integrating Institutional theory into the analysis, this study extends its application to the domain of urban planning and GIS-based sustainability initiatives. The empirical evidence confirms that reciprocal exchanges of knowledge and resources within organizations—specifically through investments in GIS adoption and support for spatial data quality—play a pivotal role in enhancing sustainable regional development. This highlights the importance of considering mediating and moderating mechanisms in understanding the relationship between technology adoption and developmental outcomes, providing a more nuanced theoretical framework than prior studies that treated GIS adoption as a direct predictor. Moreover, the positive moderating effect of institutional capacity reinforces theoretical perspectives that emphasize organizational context, governance, and technical competence as critical enablers of technology-driven sustainability. By simultaneously modeling mediation and moderation, this study contributes to the theoretical literature by illustrating how and under what conditions GIS adoption translates into sustainable development outcomes.

From a practical standpoint, the results provide clear guidance for policymakers, urban planners, and public-sector managers in metropolitan contexts like Karachi. First, the strong mediating role of spatial data quality underscores the need for organizations to invest not only in GIS technologies but also in maintaining high-quality, reliable, and up-to-date geospatial datasets. This ensures that technological adoption effectively informs planning and decision-making processes. Second, the moderating role of institutional capacity emphasizes that technology and data alone are insufficient; organizations must develop robust governance structures, technical expertise, and interdepartmental coordination to fully leverage GIS for sustainable development. Practically, this means training personnel, establishing clear protocols for data management, and fostering a culture of knowledge sharing and collaboration. Finally, by demonstrating the combined importance of GIS adoption, data quality, and institutional support, the study offers a roadmap for local authorities to systematically enhance sustainability outcomes, optimize resource allocation, and implement evidence-based urban planning strategies. Collectively, these implications reinforce the central message of the study: technology adoption, data integrity, and institutional readiness must operate in synergy to achieve meaningful sustainable regional development. The study thus bridges a gap between theory and practice, offering actionable insights for both researchers and practitioners while contributing to the broader discourse on technology-enabled sustainability in developing metropolitan contexts.

LIMITATIONS AND FUTURE RESEARCH DIRECTIONS

While the present study provides meaningful insights into the role of GIS adoption, spatial data quality, and institutional capacity in promoting sustainable regional development, several limitations should be acknowledged. First, the study employed a cross-sectional research design, which limits the ability to draw causal inferences. Although the mediation and moderation analyses provide evidence of significant relationships, longitudinal or experimental designs are needed in future research to establish causal pathways and examine changes in sustainable development outcomes over time as GIS adoption and institutional capacity evolve. Second, the data were collected from public-sector planning and development agencies in Karachi, which may limit the generalizability of the findings to other regions or contexts. While Karachi provides a relevant metropolitan setting with complex planning needs, future studies should replicate this research in other urban, rural, and international contexts to assess the robustness of the observed relationships and examine potential contextual differences in the impact of GIS adoption and data quality. Third, the study relied on self-reported survey data, which may be subject to biases such as social desirability or common method variance, despite the use of Harman's single-factor test showing no significant common method bias. Future research could incorporate objective performance indicators, such as GIS usage logs, project outcomes, or sustainability metrics, to complement self-reported perceptions and strengthen the validity of findings. Fourth, although the study examined spatial data quality as a mediator and institutional capacity as a moderator, other potentially influential factors, such as organizational culture, interdepartmental collaboration, or technology readiness, were not considered. Future research could adopt a broader multi-level framework to capture additional mediating or moderating mechanisms that may influence the relationship between GIS adoption and sustainable development outcomes. Finally, this study focused on the technical and institutional dimensions of GIS-enabled development, but did not explore the social or community engagement aspects of sustainable regional development. Subsequent research could investigate how citizen participation, stakeholder collaboration, or public awareness interact with technological adoption to enhance sustainability outcomes. Overall, addressing these limitations will provide a more comprehensive understanding of the complex dynamics linking geospatial technologies, organizational capacities, and sustainable development, thereby guiding both theory advancement and practical interventions in diverse regional and institutional contexts.

CONCLUSION

This study provides robust evidence that GIS adoption positively influences sustainable regional development, with the effect being partially mediated by spatial data quality and moderated by institutional capacity. The findings highlight that merely adopting GIS technologies is insufficient; the accuracy, reliability, and timeliness of spatial data, coupled with strong institutional support, are critical for translating technological investments into tangible sustainable development outcomes. By applying Institutional theory, the study demonstrates that organizational investment in GIS infrastructure and data quality fosters reciprocal knowledge sharing, collaboration, and commitment among employees and departments, thereby enhancing the overall capacity for sustainable regional planning. The empirical results underscore the synergistic importance of technology, data integrity, and institutional readiness, offering both theoretical and practical insights for metropolitan contexts like Karachi and similar urban settings in developing countries. Overall, the study confirms that achieving sustainable regional development requires a holistic, integrated approach where technological adoption, data quality, and institutional capacity operate in tandem. The findings contribute to the literature by providing empirical evidence on mediating and moderating mechanisms, validating theoretical propositions, and offering a roadmap for policymakers and urban planners to optimize the use of geospatial technologies for sustainable urban and regional development.

REFERENCES

Ali, A., Imran, M., Jabeen, M., Ali, Z., & Mahmood, S. A. (2023). Factors influencing integrated information management: Spatial data infrastructure in Pakistan. *Information Development*, *39*(2), 213–234.

Alzahrani, N. A., Abdullah, S. N. H. S., Adnan, N., Ariffin, K. A. Z., Mukred, M., Mohamed, I., & Wahab, S. (2024). Geographic information systems adoption model: A partial least square-structural equation modeling analysis approach. *Heliyon*, *10*(15). PMID: 39170420

Chen, S., Liu, H., Chen, J., Li, J., Jiang, N., & Ho, H. C. (2025). Community capital, gender, and depression in older caregivers for people living with dementia: A moderated mediational analysis with urban big geospatial data. *Environmental and Sustainability Indicators*, •••, 100785.

Cheng, K., Yin, J., Wang, F., & Wang, M. (2025). The impact pathway of new quality productive forces on regional green technology innovation: A spatial mediation effect based on intellectual property protection. *PLoS One*, *20*(4), e0319838. PMID: 40198691

El Gareh, F., Elmenssouri, A., Oulamine, A., & Hussain, Z. (2025). A PRISMA-Based Systematic Review on Organizational Commitment and Logistic Performance. *Knowledge Sharing and Fostering Collaborative Business Culture*, 213-240.

Hussain, A., Bhatti, M. A., Nawaz, M. A., & Ahmad, T. I. (2019). The influence of geography and its information management system on tourism planning: Mediating role of institutional factor. *PalArch's Journal of Vertebrate Palaeontology*, *16*(1), 1–16.

Hussain, Z., Khan, A., & Ali, A. (2023). The impact of user-generated content, social interactions and virtual economies on metaverse environments. *Journal of Sustainable Economics*, *1*(2), 34–44.

Khan, S. A. R., Godil, D. I., Jabbour, C. J. C., Shujaat, S., Razzaq, A., & Yu, Z. (2025). Green data analytics, blockchain technology for sustainable development, and sustainable supply chain practices: Evidence from small and medium enterprises. *Annals of Operations Research*, *350*(2), 603–627.

Lei, Y. (2024). Enhancing environmental management through big data: Spatial analysis of urban ecological governance and big data development. *Frontiers in Environmental Science*, *12*, 1358296.

Liu, Y., Hou, Y., Guo, J., & Yan, C. (2024). The impact of environment on enhancement of tourism system resilience in China: The moderating role of digital technology. *Technological Forecasting and Social Change*, *206*, 123492.

Scott, A. J. (1995). The geographic foundations of industrial performance. *Competition & Change*, *1*(1), 51–66.

Sun, Y., & Sun, H. (2021). Green innovation strategy and ambidextrous green innovation: The mediating effects of green supply chain integration. *Sustainability*, *13*(9), 4876.

Tian, R., Li, H., Jia, H., & Gu, R. (2025). The impact of digital infrastructure development on the green transformation of cities: Considering the moderating effect of human capital level. *International Review of Economics & Finance*, •••, 104695.

Wang, B., Hu, D., Hao, D., Li, M., & Wang, Y. (2021). Influence of government information on farmers' participation in rural residential environment governance: Mediating effect analysis based on moderation. *International Journal of Environmental Research and Public Health*, *18*(23), 12607. PMID: 34886344

Wang, L., Liu, L., Meng, X., Gao, Q., & Fan, M. (2024). The Relationship between Geographical Self-Efficacy and Academic Achievements in Geography: A Moderated Mediating Model. *Sustainability*, *16*(7), 2682.

Wang, X., Su, H., & Liu, X. (2025). The impact of green technological innovation on industrial structural optimization under dual-carbon targets: The role of the moderating effect of carbon emission efficiency. *Sustainability*, *17*(14), 6313.

Wei, H., Wang, Y., Yang, C., & Yu, P. (2024). Empirical Analysis of the Role of Digital Agriculture in Enabling Coordinated Development of Ecosystem Services and Human Well-Being: Evidence from Provincial Panel Data in China. *Sustainability*, *16*(23), 10199.

Yang, K. (2024). Integrating geographical information systems into urban-rural planning for equitable and sustainable urban development. *Information Development*, •••, 02666669251339133.

Yu, C., Arshad, M. A., Zhao, M., & Yao, W. (2025). How Does GIS Training Affect Turnover Intention of Highway and Bridge Industry Technicians? The Mediating Role of Career Growth and the Moderating Mechanism of Work Anxiety. *Buildings (Basel, Switzerland)*, *15*(15), 2742.

Zhang, S., Wang, J., & Guan, J. (2024). How does the energy consumption structure affect the green economic development? A spatial impact analysis. *Frontiers in Environmental Science*, *12*, 1412612.

Chapter 6
The Impact of Green Practices, Service Quality, and Tourist Motivation on Satisfaction and Revisit Intention

Elnaz Allaf Jafari
Islamic Azad University, Iran

Zohre Sadat Doaei
University of Tehran, Iran

Elahe Hosseini
https://orcid.org/0000-0003-3478-5320
Yazd University, Iran

ABSTRACT

Sustainable tourism today is recognized not only as an environmental approach but also as a competitive strategy for attracting and retaining tourists. This study aims to examine the impact of green practices, service quality, and tourist motivation on tourist satisfaction, as well as the effect of satisfaction on the intention to revisit. The statistical population consisted of domestic and international tourists who traveled to Yazd, a cultural and sustainable destination, during the spring and summer of 2025. These tourists visited the region motivated by pursuits such as culture, history, relaxation, and responsible tourism. Data were collected using a five-point Likert scale questionnaire and analyzed through structural equation modeling using PLS3

DOI: 10.4018/979-8-3373-6320-2.ch006

software. The findings revealed that green practices, service quality, and tourist motivation positively influence tourist satisfaction, which in turn plays a key role in enhancing the intention to revisit.

INTRODUCTION

Tourism is a vital and profitable sector for many developing and developed countries. With the rapid growth of tourism worldwide, there is a significant link between the tourism sector and the Sustainable Development Goals. This is because tourism development creates jobs and incomes, stimulates local businesses, and promotes responsible practices for environmental protection and infrastructure improvement (Pan & Ratasuk, 2025). The industry has consistently contributed to economic development by creating various job sectors, improving the living standards of local people, and significant growth in other related sectors. Rapid economic development and increasing personal incomes contribute to increasing demand for tourism activities worldwide (Lee et al., 2020). According to statistics from the World Tourism Organization, international tourist arrivals increased by 4% in 2019, but a significant decline was also reported in 2020-2021 due to the global pandemic. However, in the post-pandemic phase, the industry is using all strategies to strengthen the tourism sector and is performing well with open borders and tourist traffic (Morshed et al., 2022).

However, in recent decades, the attention of researchers and policymakers has shifted beyond the initial attraction of tourists to the retention and return of travelers. In this regard, the concept of revisit intention has emerged as one of the key indicators of customer loyalty in the tourism industry. Revisit intention indicates the tourist's subjective desire to return to a destination that he has previously experienced (Nejat & Fallah Lajimi, 2022). This concept not only indicates the tourist's satisfaction with the past experience, but also his desire to establish a long-term relationship with the tourist destination. In a competitive environment where multiple tourism destinations are trying to capture a larger share of the global market, a destination's ability to induce revisit intention can create a sustainable competitive advantage (Nguyen Huu et al., 2024).

Also, empirical evidence suggests that contemporary tourists, especially younger and more environmentally conscious generations, are increasingly sensitive to sustainability issues. They not only seek enjoyable tourism experiences, but also expect tourism destinations to demonstrate their social and environmental responsibility. (Schönherr & Pikkemaat, 2024) In such a situation, implementing green policies such as waste management, water and energy conservation, the use of clean transportation, biodiversity protection, and local community participation can not only help preserve natural and cultural assets, but also act as a competitive differentiator and a driver for revisit intention (Nguyen Huu et al., 2024; Allaf Jafari et al., 2025).

Meanwhile, tourism motivation, as the primary driving force of travel, plays a fundamental role in shaping tourist expectations and perceptions. Tourists who travel with motives such as seeking relaxation, discovering nature, experiencing local culture, or even an ethical commitment to the environment have different evaluation criteria for the quality and value of the experience. (Ding & Xu, 2024) On the other hand, service quality, as a tourist's perception of the match between expectations and the actual performance of service providers, is considered one of the strongest predictors of satisfaction and revisit intention. In modern tourism environments, service quality is no longer limited to room cleanliness or response speed; it also includes dimensions such as environmental responsibility and respectful interaction with the local community (Liu et al., 2025).

In the tourism literature, numerous empirical evidences have shown that the implementation of green practices in hotels and tourist destinations not only improves the destination image and customer trust (Moise et al., 2018), but also has a significant impact on tourist satisfaction and loyalty (Merli et al., 2019). Also, studies by Han et al. (2021) and Zhang et al. (2021) have consistently confirmed that high levels of tourist satisfaction significantly increase the willingness to revisit different tourism contexts. In addition, recent research such as Xia et al. (2024) also emphasizes the role of tourism motivations as a key antecedent in the formation of satisfaction. However, these findings have been mainly obtained in Western contexts or modern urban destinations and have been less generalized to cultural-indigenous destinations with inherent sustainability features such as Yazd city, which has a high potential for green tourism with its UNESCO World Heritage Site, architecture adapted to the desert climate, low-consumption lifestyle, and rich Zoroastrian culture. Despite the significant growth of tourism in Yazd and the diversity of tourists' motivations (from seeking culture and history to a desire for responsible tourism), there is still limited empirical evidence to show whether the city's service systems, including accommodations, restaurants, and tourist attractions, have been able to induce satisfaction and ultimately revisit intention by providing service quality in line with green principles and tourists' diverse motivations. This research gap highlights the necessity of examining the triple interaction of tourism motivation, service quality, and green practices in shaping tourist satisfaction and ultimately revisit intention in Yazd city.

THEORETICAL LITERATURE AND HYPOTHESIS DEVELOPMENT

Green Practices and Tourist Satisfaction

Green practices include a wide range of practices and practices aimed at minimizing environmental impacts. Also, green practices include actions designed to reduce environmental damage (Yousaf, 2021). Green practices have been highlighted in various industries for using less natural resources,reducing waste and greenhouse gas emissions, and recycling and reusing materials in the production sector. Also, due to service performance, hotels require significant consumption of energy, water, and materials, thus contributing to environmental degradation, so that the problem of green practices in hotels has been given attention (Huyen et al., 2025). In the hospitality industry, green practices are defined as the efforts or activities of hotels to carry out environmentally friendly operations to become a green hotel. Green practices in the hospitality industry require customer engagement. Therefore, in addition to the green practices criteria similar to other industries, some differentiation criteria are suggested, such as information about the hotel's good environmental practices, encouraging guests to participate in sustainable initiatives, and providing information about local environmental and cultural activities as well as public transportation options. (Merli et al., 2019) Satisfaction is considered a determinant of business success and is one of the concepts that has recently attracted the most attention in the hospitality industry in particular. Satisfaction refers to the comparison between sacrifices (costs) and rewards (benefits) that consumers perceive during and after their service experience (Moise et al., 2018). In this area, Wong et al. (2023) found that green practices all have a positive and significant effect on green customer satisfaction by studying customer behavior with a brand. In another study, Po and Jiang (2023) found that different dimensions of green practices have a positive and significant effect on perceived green value, green satisfaction, and green loyalty of hotel customers. Therefore, the following hypothesis is put forward:

> **H1: Green practices have a positive and significant effect on tourist satisfaction.**

Service Quality and Tourist Satisfaction

Liu et al. (2020) define service quality as a function of the difference between the customer's perception of the actual service and what was expected. Similarly, service quality can be defined as the outcome of an evaluation process in which the consumer compares his or her expectations with the service received. Or, it can be defined as the difference between the customer's expectations and the customer's perceptions of the service received (Khan et al., 2025). Also, satisfaction is defined as an emotional response of varying intensity that focuses on specific aspects of the acquisition or consumption and occurs when the customer evaluates the product or service. In the context of hotels, given that the service provided to customers is more than just the accommodation itself, satisfaction with a hospitality experience is defined as the sum of satisfaction with the unique elements or features of all the products and services that make up that experience (Moise et al., 2018).

On the other hand, in the field of the hotel industry, a study found that five dimensions of service quality, namely assurance, reliability, responsiveness, tangibility, and empathy, are correlated with tourist satisfaction. (Khan et al., 2025) Also, high service quality will lead to improved customer satisfaction and retention, greater market share, and consequently improved financial performance and profitability. (Abdou et al., 2022) Therefore, the following hypothesis is proposed:

H2: Service quality has a positive and significant effect on tourist satisfaction.

Tourism Motivation and Tourist Satisfaction

Academics have always studied tourist motivation as one of the most important and complex areas of tourism research. Some agree with the "Pull-Pull Theory" and believe that tourism motivation consists of two factors: "Push" and "Pull". "Push" refers to internal factors of tourists themselves, and "pull" refers to external factors related to tourism destinations. (Tang et al., 2022) Also, as a psychological concept, satisfaction is described as the pleasure and good feeling that a person experiences from receiving what he or she expects from an attractive product or service, which in fact customer satisfaction is achieved by comparing their expectations before and after consumption. Dissatisfied customers mean that customers have received less than they expected. In other words, they feel dissatisfied when they realize that the performance of the product and service does not match their expectations and does not meet them. (Abdou et al., 2022)

On the other hand, it was shown in a study that the motivation to push the tourist does not directly affect satisfaction. (Kusdibyo, 2022) But in a study, it shows a direct relationship between tourism motivation and tourist satisfaction (Deng et al., 2024). Therefore, the following hypothesis is proposed:

H3: Tourism motivation has a positive and significant effect on tourist satisfaction

Tourist Satisfaction and Revisit Intention

Revisit intention in scientific studies means the desire or intention of an individual to return to a place, service, or experience after a first visit or initial experience. This concept is studied in various fields such as tourism, marketing, and service management and is usually considered as an indicator of customer or visitor satisfaction and loyalty. (Seow et al., 2024) Also, satisfaction, as a result of psychological encounters, has considerable importance in various sectors, including tourism. This concept is usually understood as a comprehensive evaluation of the characteristics derived from the provision of products and services. In tourism, satisfaction is revealed as a mental state that is intricately intertwined with individuals' interactions with their chosen destinations. Tourist satisfaction is a central factor in shaping tourists' behavioral intentions, especially regarding their likelihood of revisiting destinations (de Sousa Saldanha & da Silva, 2024).

This strong association between satisfaction and behavioral intentions forms an important foundation for predicting tourist behavior. Recent empirical research, including Han et al. (2021), found a strong positive correlation between tourist satisfaction and revisit intention to cultural heritage destinations, which emphasizes the importance of managing visitors' experiences to enhance satisfaction. Similarly, Zhang and Yi Man Li (2021) showed that satisfaction with sustainable tourism initiatives has a positive effect on tourists' intention to support such practices, suggesting that satisfaction-based strategies can foster sustainable tourism development. (Zhang et al., 2021) In another study by Hashish et al. (2022), it was proven that the intention to revisit green hotels and the intention to pay extra for accommodation in eco-friendly hotels are positively influenced by green satisfaction. (Merli et al., 2019)

Therefore, the following hypothesis is put forward:

H4: Tourist satisfaction has a positive and significant effect on revisit intention.

The conceptual model of the research has been selected as follows, based on the research literature.

Figure 1. Conceptual model (based on the theoretical foundations of research)

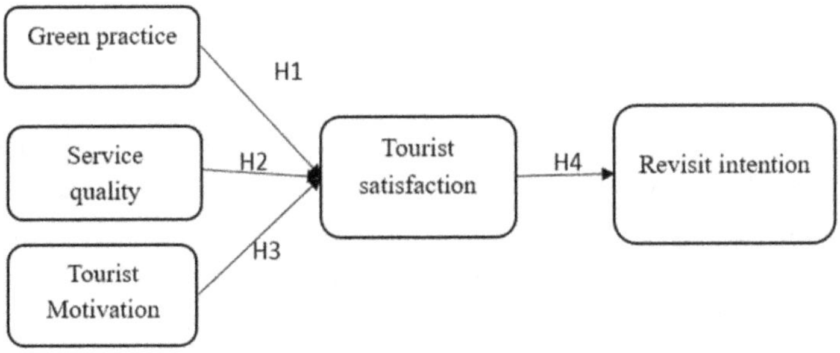

RESEARCH METHOD

The present study is applied in terms of purpose and is considered a descriptive-correlational research in terms of method. The statistical population of the study includes those domestic and foreign tourists who have traveled to Yazd as a tourist destination in a specific period of time. Among them, tourists with various motives including searching for culture, history, relaxation and responsible tourism have been considered. The statistical population includes all tourists visiting Yazd in the spring and summer of 1404. Based on the estimate and considering the high volume of tourists, the sample size was determined using the Morgan table and sampling formulas and 386 tourists were selected from among the tourists through simple random sampling. The sample selection criteria were registration of a trip to Yazd and readiness to respond to questionnaires related to tourism motives, service quality and green practices. A questionnaire was used to collect data. In this questionnaire, a 5-point Likert scale (1- strongly disagree, 2- disagree, 3- no opinion, 4- agree and 5- strongly agree) was used to measure the variables of the conceptual model. This study examined the relationships between variables using PLS3 software and the structural equation modeling method. The most important reason for using this software is that the distribution does not require normality compared to other existing software (Kline, 2015). Also, using the PLS method is compatible with constructive measurement models. It is applicable for reflective and developmental models. Similarly, if the data is normal or the sample is large, partial least squares can be used (Rousta & Allaf Jafari, 2024).

Measurement, reliability, validity

In this study, the variables of tourism motivation, service quality and green practices, independent, intention to revisit the dependent variable, tourist satisfaction were considered as mediator variables. In order to ensure the accuracy and correctness of the research results, the technical characteristics of the questionnaire were evaluated in two parts: validity and reliability using different criteria. The validity of the questionnaire was evaluated and confirmed through content and construct validity. To measure the content validity of the tool, the opinions of professors and experts knowledgeable in the subject were used, and to measure the construct validity, the structural equation model of convergent and divergent validity was used. In order to determine the convergent validity, the average standard deviation index (AVE) was used, and for the divergent measurement, the extracted root mean square index was used. Cronbach's alpha and combined reliability were used to measure reliability. Cronbach's alpha coefficient is a factor whose value varies from 0 to 1, a Cronbach's alpha value higher than 0.7 indicates acceptable reliability. Composite composite reliability, which has the advantage over Cronbach's alpha in that the reliability of the constructs is calculated not in absolute terms but according to the correlation of their constructs with each other. If the composite reliability value for each construct is above 0.7, it indicates appropriate internal reliability for the measurement models, and a value less than 0.6 indicates the absence of reliability (Hosseini et al., 2024).

Table 1. Factor loadings, composite reliability, cronbach's alpha, and convergent validity of the model

Variables	Index	Factor Loading	Cronbach's Alpha	AVE	Composite Reliability
green practice	q1	0.756	0.749	0.570	0.841
	q2	0.750			
	q3	0.740			
	q4	0.773			
service quality	q5	0.739	0.780	0.531	0.850
	q6	0.724			
	q7	0.711			
	q8	0.736			
	q9	0.735			

continued on following page

Table 1. Continued

Variables	Index	Factor Loading	Cronbach's Alpha	AVE	Composite Reliability
Tourist Motivation	q10	0.740	0.794	0.548	0.858
	q11	0.762			
	q12	0.716			
	q13	0.742			
	q14	0.740			
tourist satisfaction	q15	0.774	0.785	0.539	0.854
	q16	0.710			
	q17	0.757			
	q18	0.710			
	q19	0.717			
Revisit intention	q20	0.738	0.795	0.549	0.859
	q21	0.744			
	q22	0.741			
	q23	0.736			
	q24	0.745			

As Table 1 shows, the values of the extracted root mean square variance are greater than the minimum acceptable value of 0.5; therefore, the research variables have divergent validity. Also, considering that the values of the extracted root mean square variance are greater than the correlation of the variable in question with other variables, divergent validity is acceptable if the numbers in the main diagonal are greater than their underlying values. So, it can be said that the variables have validity and their divergent validity is also confirmed (Hair et al,.2021).

Table 2. Divergent validity of the model

	Revisit intention	Tourist Motivation	green practice	service quality	tourist satisfaction
Revisit intention	0.741				
Tourist Motivation	0.703	0.740			
green practice	0.715	0.728	0.755		
service quality	0.725	0.713	0.718	0.729	
tourist satisfaction	0.731	0.711	0.703	0.726	0.734

Based on the above-mentioned materials and the results obtained from the output of the SmartPLS 3 software in Tables 1 and 2, the measurement models have appropriate validity (convergent and divergent) and reliability (factor loading, composite reliability coefficient and Cronbach's alpha).

FINDINGS

In order to examine the fit of the model at three levels of measurement, structural and general, several criteria are used to examine the fit of the structural model of the research using the partial least squares method, the first and most basic criterion being the significance coefficients or the values of the t-statistics. The fit of the structural model using t-coefficients is such that these coefficients must be greater than 1.96 in order to confirm their significance at a 95% confidence level. The results showed that the values obtained for the path of the effect of green practices on tourist satisfaction were (3.620), the path of the effect of service quality on tourist satisfaction was (4.938), the path of the effect of tourism motivation on tourist satisfaction was (6.353), and the path of the effect of tourist satisfaction on revisit intention was (22.255), respectively, and the critical value of the four paths was greater than the critical value (1.96) at a confidence level of 95 percent. This indicates the significance of the paths, the appropriateness of the structural model, and the confirmation of the four research hypotheses (Figure 2).

Figure 2. T-statistics

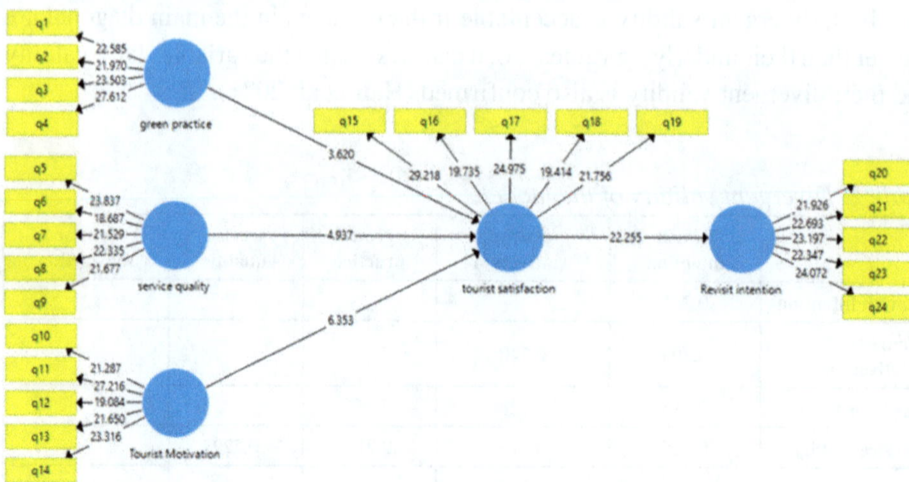

The coefficient of determination criterion (R^2)

The second criterion for examining the fit of a structural model in a study is the R^2 coefficients related to the endogenous latent variables of the model. This criterion is used to examine the intensity of relationships between structures, which is only applicable to dependent and not independent structures. R^2 is a criterion that indicates the effect of exogenous variables on an endogenous variable, and three values of 0.19, 0.33 and 0.67 are considered for weak, medium and strong R^2 values. (Hair et al., 2021) In this study, the coefficient of determination of the psychological empowerment variable is (0.408) and the value of this coefficient for the revisit intention variable is (0.625), which indicates the appropriateness of the level of this effect, but the coefficient of determination for tourist satisfaction is (0.725), which indicates the strength of the level of this effect. Also, according to Figure 3, the results have been shown.

Figure 3. Model in the case of standard factor loading coefficients

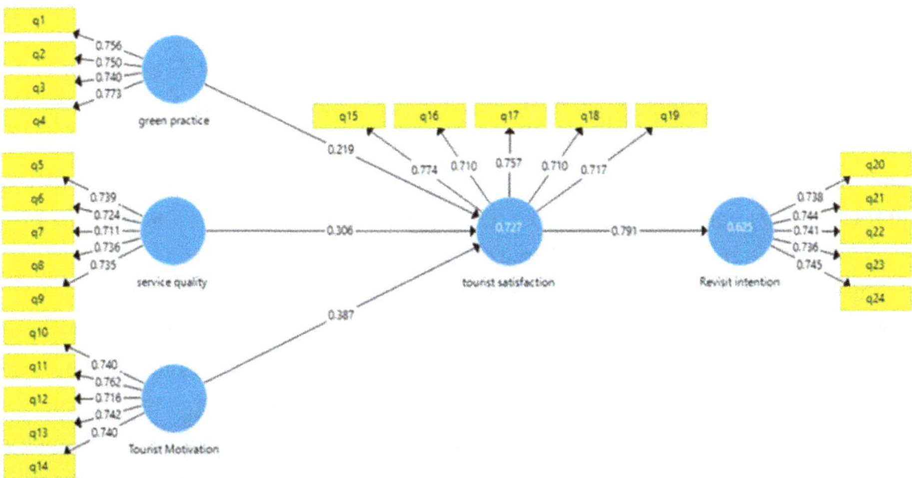

The results also showed that the impact coefficient for the impact coefficient of green practices on tourist satisfaction is (0.219), the impact coefficient of service quality on tourist satisfaction is (0.306), the impact coefficient of tourism motivation on tourist satisfaction is (0.387), and the impact coefficient of tourist satisfaction on revisit intention is (0.791). Figure 3 shows that most of the impact coefficient of 0.791 is related to the fourth hypothesis.

Q^2 criterion

The Q^2 criterion is calculated for all dependent constructs and represents the product of the combined values of the research constructs by their respective coefficient of determination values. This criterion, introduced by Stone and Geiser (1975), determines the predictive power of the model in dependent variables. According to them, models that have acceptable structural part fit should be able to predict the indicators related to the endogenous constructs of the model. This means that if the relationships between the constructs in a model are defined correctly, the constructs will be able to have a sufficient impact on each other's indicators and thus the hypotheses will be confirmed correctly. The value should be 0.02, 0.15 and 0.35 for all endogenous constructs as low, medium and strong predictive power (Kline, 2015). This criterion in the present study is for tourist satisfaction (0.367) and revisit intention (0.324), which indicates an acceptable fit of the structural model of the research.

General Model Fit

The general model of the research is examined using the GOF criterion. This criterion is the geometric mean of the average coefficient of multiple determination in the average commonality. Three values of 0.01, 0.25 and 0.36 are introduced as weak, medium and strong values for GOF. Given that the average commonality is equal to (0.547) and the average coefficient of determination is equal to (0.674).

$$GOF = \sqrt{\overline{Communality} \times \overline{R^2}} = 0.674$$

The GOF value obtained indicates a strong fit of the model.

According to the research hypotheses, structural fit using t-coefficients is as follows: these coefficients must be greater than 1.96 in order to confirm their significance at a confidence level of 0.95. Of course, it should be noted that t-numbers only indicate the accuracy of the relationships and the intensity of the relationship between the structures cannot be measured with them, and path coefficients also indicate the positive or negative effect of one variable on another variable.

Table 5. The results of the hypothesis test

Row	Research Hypothesis	Impact Factor	T -Statistics	Result
1	**Tourist Motivation -> tourist satisfaction**	0.387	6.353	Confirmed
2	**green practice -> tourist satisfaction**	0.219	3.620	Confirmed
3	**service quality -> tourist satisfaction**	0.306	4.937	Confirmed
4	**tourist satisfaction -> Revisit intention**	0.791	22.255	Confirmed

As can be seen from Table 4, all research hypotheses have a critical value greater than 1.96 and have been confirmed.

DISCUSSION AND CONCLUSION

This study demonstrated that the success of sustainable tourism in Yazd city requires the integration of three key elements: green practices, which are not only an environmental obligation but also an integral part of the cultural identity and native nature of the destination; service quality, which goes beyond technical criteria to include cultural dimensions, indigenous authenticity, and environmental responsibility; and a deep understanding of tourists' motivations considering their individual and group differences. The findings emphasize that these three factors collectively lead to a profound and lasting increase in tourist satisfaction, which acts as a strategic bridge between the tourist experience and future behavior, including the intention to revisit. Additionally, reinforcing these factors by leveraging Yazd's native and cultural approaches can contribute to preserving its unique heritage, enhancing tourist loyalty, and fostering responsible and sustainable tourism development. Consequently, successful tourism in Yazd is based on creating a meaningful experience aligned with cultural and environmental values, emphasizing long-term relationships and tourist loyalty instead of merely attracting mass tourism.

Hypothesis 1: Green practices have a positive and significant impact on tourist satisfaction. Specifically, Moise et al. (2018) demonstrated that implementing green practices in hotels not only strengthens trust and a positive image of the destination but also directly affects guest satisfaction. Similarly, Merli et al. (2019) emphasized that tourists witnessing sustainable actions (like water conservation, recycling, and use of local products) perceive the hotel's ethical responsibility and consequently report higher satisfaction. These findings are supported by Wong et al. (2023) and Po & Jiang (2023), who showed that green practices improve tourist satisfaction by increasing perceived green value.

In the context of Yazd, this relationship gains special significance. Yazd, with its UNESCO World Heritage status, indigenous architecture adapted to desert climate (such as windcatchers, qanats, and adobe houses), traditional water management systems, and the frugal lifestyle of the local community, possesses inherent and cultural green practices rooted in centuries of ecological harmony. These features can be understood as "native green practices" that not only align with sustainable tourism principles but also resonate deeply with tourists who travel motivated by ethical, cultural, or environmental reasons. Therefore, strengthening and integrating these native practices as part of the tourism experience can significantly enhance tourist satisfaction.

It is recommended that tourism service providers, including hotels, restaurants, and tour agencies, communicate these practices to tourists through the cultural and historical narratives of Yazd. For example, the use of plastic-free services should not be merely regarded as an environmental policy but as a continuation of the ancient tradition of frugality and indigenous wisdom, like the qanat system or adobe architecture. Local tourism organizations can support tourism businesses by designing specialized training programs to embed these "native green practices" into the tourist experience. Meanwhile, policymakers should incentivize responsible and identity-centric tourism by supporting traditional and indigenous accommodations inherently aligned with sustainability principles and by granting official "Yazd Green Tourism" labels to committed units.

Hypothesis 2: Service quality has a positive and significant impact on tourist satisfaction. This hypothesis aligns with findings by Liu et al. (2020), Khan et al. (2025), and Abdou et al. (2022), who identified service quality as one of the strongest predictors of tourist satisfaction in the tourism industry. However, this study highlights that in cultural and indigenous destinations like Yazd, the concept of service quality transcends traditional criteria such as cleanliness, responsiveness, or tangibles and includes dimensions such as cultural quality, indigenous authenticity, and environmental responsibility. For instance, providing educational explanations about windcatcher architecture, using local products in restaurants, or encouraging guests to participate in cultural heritage preservation programs are not only part of the unique tourism experience in Yazd but are also indicators of service quality in this specific context. This approach allows tourists to feel not only technically satisfied but also deeply connected to the destination's identity, ultimately enhancing their qualitative and quantitative satisfaction.

Therefore, it is recommended that tourism service providers in Yazd expand their quality frameworks beyond operational dimensions into cultural and sustainability discourses. This can be achieved through training staff about local history and architecture, designing interactive experiences based on cultural heritage, and integrating green principles in all tourist touchpoints from reservation to departure. Additionally, local tourism organizations can motivate service units by developing "cultural-green quality" standards and awarding specialized certifications to prioritize these richer service dimensions.

Hypothesis 3: Tourist motivation has a positive and significant impact on tourist satisfaction. This hypothesis is consistent with recent studies such as Xia et al. (2024), Deng et al. (2024), and Ding & Xu (2024), which emphasized the key role of intrinsic motivations (like the need for relaxation, escape from daily routine, or search for meaning) and extrinsic motivations (such as cultural, natural, or historical attractions of the destination) in shaping expectations and ultimately tourist satisfaction. In Yazd's specific context, these motivations acquire unique dimensions;

tourists might be influenced by desires to "experience Zoroastrian culture," "explore desert-adapted indigenous architecture," "familiarize with qanat systems," or even "escape the hustle of modern cities." These motivations directly determine tourists' criteria in assessing the quality of experience. For example, tourists seeking relaxation are sensitive to quiet and natural spaces, whereas culturally motivated tourists value deeper information about local heritage. Hence, service systems from accommodations and restaurants to tourist attractions and local guides must be designed and personalized based on this motivational diversity to effectively meet tourists' mental expectations, resulting in deeper and more sustainable satisfaction.

To achieve this, it is recommended to use tools such as brief questionnaires during reservation or tourist arrival to identify primary travel motivations and provide customized experiences accordingly. Furthermore, training tour guides to recognize and respond to different tourist motivations and designing targeted tour packages (such as "Zoroastrian Culture Tour" or "Yazd Indigenous Life Experience") can significantly enhance tourist satisfaction and loyalty.

Hypothesis 4: Tourist satisfaction has a positive and significant impact on revisit intention. In this regard, Han et al. (2021) demonstrated that in cultural heritage destinations, tourist satisfaction is a strong predictor of return intentions. Zhang et al. (2021) emphasized that satisfaction derived from sustainable tourism initiatives not only strengthens loyalty but also encourages active support for the destination by tourists. These findings are further supported by Hashish et al. (2022) and de Sousa Saldanha & da Silva (2024), documenting a direct and strong relationship between satisfaction and behavioral intentions, including revisit intentions, across different tourism destinations.

In Yazd's context, this relationship becomes particularly critical; tourists who have had satisfying experiences—intelligently combining native green practices, cultural-environmental service quality, and alignment with their personal motivations—are not only more inclined to return but also serve as unofficial ambassadors, attracting like-minded tourists through positive word-of-mouth. Therefore, satisfaction in Yazd is not only a short-term outcome but a driving force for long-term and sustainable tourism.

To reinforce this relationship, it is proposed that Yazd's destination managers establish post-trip follow-up systems (such as digital surveys or loyalty programs) to maintain communication with tourists after departure. Additionally, designing "special return programs" (like discounts for second visits or invitations to local cultural festivals) can convert revisit intentions into actual behavior. Ultimately, investing in creating deep, meaningful experiences aligned with tourist values will not only increase satisfaction but also position Yazd as a destination to which tourists return repeatedly.

In conclusion, this research indicates that the success of sustainable tourism in Yazd requires integrating three key elements: first, green practices should be considered not merely operational requirements but inseparable parts of the destination's cultural and environmental identity; second, service quality in this specific context should transcend technical standards and include cultural dimensions, indigenous authenticity, and environmental responsibility; and third, a profound understanding of tourists' motivations—whether those seeking cultural and historical experiences or those traveling with ethical and sustainability motivations—to offer services that not only meet expectations but create a sense of value alignment and deeper satisfaction. The combination of these three elements not only helps preserve Yazd's unique heritage but also fosters tourist loyalty and repeat visits as indicators of genuine sustainable tourism.

Within this framework, tourist satisfaction is not an end goal but a strategic bridge between the experience gained at the destination and future tourist behavior, especially revisit intentions. This perspective emphasizes that successful tourism occurs only when a deep, meaningful experience aligned with individual and collective values is created. Such an approach not only preserves Yazd's cultural and environmental heritage but is also economically more sustainable and efficient compared to traditional models based on mass tourism attraction, as it focuses on building long-term relationships, loyalty, and conscious tourist return rather than dependence on short-term tourist flows.

REFERENCES

Abdou, A. H., Mohamed, S. A. K., Khalil, A. A. F., Albakhit, A. I., & Alarjani, A. J. N. (2022). Modeling the relationship between perceived service quality, tourist satisfaction, and tourists' behavioral intentions amid COVID-19 pandemic: Evidence of yoga tourists' perspectives. *Frontiers in Psychology, 13*, 1003650. PMID: 36211929

Allaf Jafari, E., Asayesh, F., & Rostamian, M. (2025). Towards building a sustainable future with the tech revolution. *International Journal of Human Capital in Urban Management, 10*(2), 317–328.

De Sousa Saldanha, E., & da Silva, H. B. (2024). Linking Tourists' Motivation to Behavior Intentions through Government Policies and Tourists' Satisfaction: Evidence from Ramelau Mountain as a Tourist Destination in Timor-Leste. *Timor Leste Journal of Business and Management, 6*, 12–25.

Deng, Y., Wang, X., & Ma, C. (2024). The impact of sports tourism motivation on tourist loyalty: The chain mediation effect of experience quality and tourist satisfaction. *International Journal of Tourism Research, 26*(5), e2769.

Ding, X., & Xu, J. (2024). Urban professionals' restorative tourism: Exploring the role of perceived environmental restorativeness, push and pull motivations and destination attributes on tourism expectations. *Frontiers in Psychology, 15*, 1293050. PMID: 38566934

Hair, J. F.Jr, Hult, G. T. M., Ringle, C. M., Sarstedt, M., Danks, N. P., & Ray, S. (2021). *Partial least squares structural equation modeling (PLS-SEM) using R: A workbook*. Springer Nature.

Han, J., Zuo, Y., Law, R., Chen, S., & Zhang, M. (2021). Service quality in tourism public health: Trust, satisfaction, and loyalty. *Frontiers in Psychology, 12*, 731279. PMID: 34526941

Hosseini, E., Doaei, Z. S., Jamadi, A., & Yazdani, M. (2024). Examining the effect of resilience on job engagement with the mediation of psychological empowerment and job burnout. *Interdisciplinary Journal of Management Studies, 17*(2), 475–490.

Huyen, N. N. T., Hong, H. D., & Thi, L. H. (2025). Green practices: Building green image and green trust for green revisit intentions in the hospitality industry. *Journal of Open Innovation, 11*(1), 100481.

Khan, F., Yusoff, R. M., & Kakar, P. (2025). Impact of service quality and customer satisfaction on tourism industry. *Journal of Advanced Research in Social and Behavioural Sciences, 39*(1), 1–11.

Kline, R. B. (2015). *Principles and practice of structural equation modeling*. Guilford publications.

Kusdibyo, L. (2022). Tourist loyalty to hot springs destination: the role of tourist motivation, destination image, and tourist satisfaction. *Leisure/Loisir, 46*(3), 381-408.

Lee, M.-T., Liu, J.-M., & Borazon, E. Q. (2020). Evaluating the effect of perceived value of ecosystem services on tourists' behavioral intentions for Aogu Coastal Wetland. *Sustainability, 12*(15), 6214.

Liu, J., Kuai, X., & Wang, P. (2025). Research on the tourism service quality evaluation of Gongbei Lingnan community under the perspective of SERVQUAL theory. *Frontiers in Sustainable Cities, 6*, 1502234.

Liu, Y., Huang, D., Wang, M., & Wang, Y. (2020). How do service quality, value, pleasure, and satisfaction create loyalty to smart dockless bike-sharing systems? *Revista Brasileira de Gestão de Negócios, 22*(03), 705–728.

Merli, R., Preziosi, M., Acampora, A., & Ali, F. (2019). Why should hotels go green? Insights from guests experience in green hotels. *International Journal of Hospitality Management, 81*, 169–179.

Moise, M.-S., Gil-Saura, I., & Ruiz-Molina, M.-E. (2018). Effects of green practices on guest satisfaction and loyalty. *European Journal of Tourism Research, 20*(20), 92–104.

Morshed, M. M., Hashim, R. A., & Mahmood, R. (2022). Tourist Satisfaction, Destination Image, Loyalty and Revisit Intention to Malaysia's National Parks. *The Asian Journal of Professional & Business Studies, 3*(2).

Nejat, S., & Fallah Lajimi, H. (2022). Identifying and prioritizing Factors Affecting the Revisit Intention of the Destination Brand in Iran with the Best-Worst Method. *Journal of International Marketing Modeling, 3*(1), 62–84.

Nguyen Huu, T., Nguyen Ngoc, H., Nguyen Dai, L., Nguyen Thi Thu, D., Truc, L. N., & Nguyen Trong, L. (2024). Effect of tourist satisfaction on revisit intention in Can Tho City, Vietnam. *Cogent Business & Management, 11*(1), 2322779.

Pan, C., & Ratasuk, A. (2025). Determinants of Revisit Intention: A Case of Snorkeling Tourism in Hainan Province. *Journal of ASEAN PLUS Studies, 6*(1), 43–64.

Po, S., & Jiang, M. (2023). The effects of green practices on green customer loyalty in the hotel industry: Multi-mediating role of green perceived value and green satisfaction. *International Journal of Research, 11*(10), 95–106.

Rousta, A., & Allaf Jafari, E. (2024). Impact of environmental knowledge, responsibility and concern on sustainable consumption behavior: Does customer attitude matter? *Management of Environmental Quality*, *35*(8), 1858–1877.

Schönherr, S., & Pikkemaat, B. (2024). Young peoples' environmentally sustainable tourism attitude and responsible behavioral intention. *Tourism Review*, *79*(4), 939–952.

Seow, A. N., Foroughi, B., & Choong, Y. O. (2024). Tourists' satisfaction, experience, and revisit intention for wellness tourism: E word-of-mouth as the mediator. *SAGE Open*, *14*(3), 21582440241274049.

Tang, H., Wang, R., Jin, X., & Zhang, Z. (2022). The effects of motivation, destination image and satisfaction on rural tourism tourists' willingness to revisit. *Sustainability*, *14*(19), 11938.

Wong, C. B., Law, M., & Wu, W. C. B. (2023). Determinants of customer loyalty: A green marketing perspective. *10th International Conference on Business, Accounting, Finance and Economics*

Xia, L., Lee, T. J., & Kim, D. K. (2024). Relationships between motivation, service quality, tourist satisfaction, quality of life, and spa and wellness tourism. *International Journal of Tourism Research*, *26*(1), e2624.

Yousaf, Z. (2021). Go for green: green innovation through green dynamic capabilities: accessing the mediating role of green practices and green value co-creation. *Environmental Science and Pollution Research International*, *28*(39), 54863–54875. PMID: 34018115

Zhang, Y., Shao, W., & Thaichon, P. (2021). Investigating tourist post-travel evaluation and behavioural intention: A cultural intelligence perspective. *Asia Pacific Journal of Marketing and Logistics*, *33*(10), 2037–2053.

Chapter 7
Leveraging Geospatial Information Systems for Sustainable Development

Yamijala Suryanarayana Murthy
https://orcid.org/0000-0002-9561-5395
KL Business School, KL University, India

Rohit Bansal
https://orcid.org/0000-0001-7072-5005
Rockford College, Sydney, Australia

Ridhima Sharma
Vivekananda Institute of Professional Studies-TC, India

Zahid Hussain
https://orcid.org/0009-0006-2182-9303
KASBIT University, Pakistan

Nishita Pruthi
https://orcid.org/0000-0002-6094-0972
Asian School of Business, India

ABSTRACT

Geospatial Information System (GIS) has come as a radical force in handling multidimensional sustainability issues within the environmental, social and economic spheres. GIS offers strong analytical to aid the planning, monitoring, and evaluation of evidence-based decisions in the countries that face the issues of climate change, increasing urbanization, resource depletion, and uneven socio-economic conditions. The role of GIS within sustainable development is considered in this chapter in an

DOI: 10.4018/979-8-3373-6320-2.ch007

integrative manner in that it encompasses the theoretical background, empirical studies (2010-2025) and methodological aspects. It summarizes the developments in spatial analytics, remote sensing, machine learning integration, and geospatial big data management and clarifies the importance of the discussed approaches to land-use planning, disaster risk reduction, environmental conservation, renewable energy, and sustainable urban development.

INTRODUCTION

With the global societies experiencing uncertainties in their ecological systems, economic difficulties, and more susceptible individuals, sustainable development has emerged as the primary international issue. The Sustainable Development Goals (SDGs) by the United Nations are a global agenda that requires systematic and evidence-based approaches to proper execution. Geographic information systems (GIS) have become part of this intent in recent years, enabling the capture, analysis, visualization and interpretation of spatial data in order to enable the sustainability-oriented decisions. The improvement of artificial intelligence, big data analytics, cloud computing, and remote sensing increased the growth of GIS. The innovations have allowed governments, researchers, and industry professionals to solve complex sustainability challenges, including land-use conflict, calamities, and associated climate change, biodiversity loss, and resource shortage (Goodchild, 2018; Li et al., 2020). Moreover, GIS has the ability to deliver spatial intelligence that is important to create resilient urban centers, transportation networks, and conservation plans (Kabisch et al., 2021). The available literature refers to GIS as a sustainability driver because it can integrate multidisciplinary data and shows the trends that have not been realized, when the more traditional approach has been employed.

GIS frameworks also contribute to participatory governance and bring the extent of transparency and community participation to the process of environmental decision-making (Brown and Kytta, 2018). However, the use of GIS whether it has extensive literature or not does not remove research gaps. Much of the literature is highly domain specific and very little is known about holistic, cross-sectoral, GIS system sustainability. The application of GIS in order to support the integration of the environmental, social, and economic factors, in general, is not explored in numerous studies. In addition to this, the scholars concentrate on technical applications, but seldom comment on organizational, ethical, and CEO implication of GIS adoption, yet another gap is also evident in the research done on technology-integration at present such as the base paper in the appendix. This chapter is devoted to the attempt to fill all these gaps by a wide-interdisciplinary synthesis of GIS to sustainable development. The chapter provides a critical evaluation of the theoretical base, approaches, uses of GIS in strategic management and future opportunities of the GIS in the sustainability environment using evidence since the year 2010 to 2025. It is significant as it gives a broad perspective on the issue that can be applied to empower scholars, policymakers, urban planners and sustainability practitioners to use GIS to enhance socio-environmental resiliency and informative governance.

LITERATURE REVIEW

Early research placed GIS in the core of technical means of land-use mapping and land-use change, and hydrological analyses (Longley et al., 2011). These pioneering works provided the foundation of further studies that have put the use of GIS in multidimensional areas where spatial analytics took the center stage in the comprehension of urban processes, environmental dynamics, and spatial distribution of socio-economic inefficient. With the rise in international interest in sustainability, especially with the introduction of the Sustainable Development Goals (SDGs), GIS became a more and more important tool to quantify, visualize, and monitor progress on global and regional sustainability related issues (Vinuesa et al., 2020). One of the most studied factors of GIS studies has been environmental sustainability. Remote sensing and geospatial analytics have been used by scholars to identify deforestation, soil degradation, biodiversity loss, and water stress nearly in real-time (Kumar et al., 2014). New technologies of remote sensing have also allowed to analyze vegetation indices, evapotranspiration, carbon sequestration, and land-cover changes at various scales of space and time (Coops & Wulder, 2019). Later researches that incorporate machine learning into GIS processes have enhanced predictive validity of environmental risk modeling especially of natural hazards like landslides, floods, wildfires, and coastal erosion (Pham et al., 2022). Such a shift in descriptive to predictive geospatial intelligence is a larger one within the sustainability science literature on anticipatory governance and proactive risk reduction, similar to the organizational changes driven by technology mentioned in the base chapter. The study of urban sustainability also indicates the growing power of GIS in dealing with the growing urbanization processes, strain in cities infrastructures, and risks associated with climate change in cities. Smart city planning, optimization of public transit networks, and locations of interventions to curb the effects of urban heat islands have been supported with the help of GIS-based urban models (Yeh et al., 2017). GIS in conjunction with multicriteria decision analysis (MCDA) has presented with evidence-based methodologies of locating renewable energy installations, green open spaces, as well as waste management facilities, thus aiding resource-efficient and climate-resistant urban development (Ali and Mohammed, 2021). Moreover, fast growing cities in the Global South have found geospatial tools to be instrumental in mapping informal settlements, evaluating the spatial vulnerability, and enhancing service delivery (Hidayati et al., 2020). These lessons are consistent with the way organizational and policy research is calling upon data-driven, responsive systems that can address a variety of community needs and sustainability demands- the same way agility in the system is crucial in your base paper. Other than the fields of environment and urban, GIS has been a key element in the study of socio-economic sustainability. GIS has been used in studies of poverty mapping, demographic change, the distribution of public health, and service distribution to identify hidden spatial inequities that are common in aggregate statistics (Musa et al., 2015). The recent development of volunteered geographic information (VGI) and participatory GIS has further democratized the production of spatial data by therefore enabling communities to add their local knowledge to development planning and governance of natural resources (Brown and Kytta, 2018). In the COVID-19 pandemic, GIS was given a new role because it allowed spatial modeling of the rate of infection, access to health care, and vulnerability of the population, and proved to be useful in crisis management and resilience to health services (Fang et al., 2021). Since 2018, the interplay between GIS and artificial intelligence, big data analytics, cloud computing, and sophisticated remote sensing has tremendously increased the amount and the size of geospatial applications. GIS models based on AI have improved the predictive analytics of climate adaptation, transportation optimization, and land-use planning by processing large geospatial datasets in real-time (Zhao et al., 2023). Cloud computing including Google Earth Engine has lowered the computational hurdles therefore enabling researchers and practitioners worldwide to carry out vast environmental evaluation and sustainability surveillance without sophisticated local systems. Such technological developments reflect the larger trends of digital transformation observed in the base chapter, especially the input of intelligent systems into the process of decision-making and governing.

Research Gap

Even though geospatial technologies have greatly improved and their contribution to sustainability research is on the rise, there exist a number of gaps that are critical in the current body of knowledge. Much of existing literature is specific to a particular domain, be it environmental monitoring, urban planning, or disaster management per se, and little has been said about how GIS can comprehensively incorporate environmental, social, and economic aspects of sustainability in a single analytical process. Although much of the literature is based on technical aspects of GIS, very little explores organizational, managerial, and governance mechanisms needed to facilitate successful geospatial application in practice of policy and planning matters. There also exists a gap in solid empirical data of the effects of GIS based decision making on the long term sustainability dynamics especially in the resource-strained and developing world where information paucity, technical incompetence and poor infrastructures act as stumbling blocks to its implementation. Also, the issues of data privacy, ethical geospatial action, and digital inequity are still underresearched, despite the growing use of high-resolution imagery, AI-powered analytics, and citizen-created geospatial data in GIS.

THEORETICAL FOUNDATION

Spatial Decision Support Theory

The SDS theory provides a paradigm to the understanding of GIS enhancement of decision-making in complex sustainability situations. According to Malczewski (2010), incorporating spatially explicit data into the framework of the decision-making process will help to make the decision making more rational, clear, and evidence-based. The theory emphasizes that many sustainability issues including land-use controversies, environmental degradation, and allocation of resources are inherently spatial and require decision-making frameworks to take into account geographic differences, interactions that occur in space, and location-related restrictions. This theory is operationalized through GIS that combines various data such as the environmental, demographic, economic and infrastructural data. It uses multicriteria evaluation techniques to help planners to determine the merits and demerits of different sustainability goals. When planners decide on locating renewable energy, conservation and urban development, the planners need to take into account different factors, which can be the suitability of the land to the project, environmental sensitivity and the economic impact of the project. Spatial Decision Support Theory considers GIS as an important analytical tool to assist in performing complex, multi-criteria sustainability assessments.

Systems Theory and Sustainable Development

The Systems Theory describes sustainable development as the dynamic process of the interaction of ecological, social, and economic subsystems, which is an important framework. According to Meadows (2011), sustainability concerns take place as a result of interactions, feedback loops and interdependence between natural systems, activities of human action and institutional structures. GIS visualises the interrelationships both between layers and within an entity and therefore facilitates systems thinking. Changes in one subsystem can be indicated by their relationship to other subsystems as administration of land use, population, ecosystem services, and climate hazards can be displayed on the same GIS. This capability allows policymakers to observe the effect of decisions on development to the entire system. This is an example of how urbanization is affecting water supply or how the high rate of agriculture is affecting biodiversity. GIS assists in making a global plan using systems theory concept such as integration, interconnection and feedback-driven decision-making. This is possible by demonstrating the relationship between sizes. This implies that GIS is not a map-making tool. It also resorts to systems thinking into place-specific sustainability plans.

Socio-Technical Systems Framework

Socio-Technical Systems (STS) framework will contribute to the GIS utilization with a more holistic view of people and organisations, thus strengthening the theoretical basis on which the concept is based. According to this theory, technology systems, such as GIS, do not operate in isolation but rather in complex social, institutional and ethical contexts. Geospatial deployment does not need just sophisticated technology. It must also have good people, good regulations, good administration and good work culture. The introductory chapter has already portrayed the fact that to be lasting technology integration requires digital tools to correspond to leadership skills, change management practices, and governance frameworks that bring everybody to the table. Other organisations apply GIS to serve the environment whereas others are facing issues such as poor data management, insufficient training, resistance to change, and ethical issues relating to privacy and monitoring. It is possible to explain why this is occurring with the help of STS thinking. The paradigm highlights how socio-organizational enablers (e.g. stakeholder participation, cross-departmental cooperation, clear data management, and capacity building efforts) should be used to complement the effectiveness of the geospatial technologies. The STS approach demonstrates that sustainability solutions implemented using GIS require a high level of technology and social and institutional willingness.

Technology Acceptance and Diffusion Theories

The realization of sustainable development with the help of GIS can be viewed in the prism of concepts regarding the mechanism of the adoption and diffusion of new technologies. The Unified Theory of Technology Acceptance (UTAUT) assumes that the perceived utility, ease of use, social influence, and facilitating conditions are the elements that determine the utilization of digital tools (Venkatesh et al., 2012). In the case of GIS, the stakeholders are more ready to use geospatial systems in planning and management when they can realize the visible pay-offs such as having more accurate data, faster visualization of the data or better decisions. The Diffusion of Innovation Theory also makes us see why GIS diffuses at a higher or a slower rate over different regions and fields. It achieves this through examining issues such as the compatibility of the system with other systems, ease of use, complexity of the technology, and ease of vision. These theories are useful in understanding why localities that have less resources or institutional support are more likely to embrace GIS gradually compared to localities that have more resources, including developed economies and government bodies. Another area that they emphasize is the importance of training individuals, demonstrating the use of geospatial technologies, developing pilot programs, and receiving the assistance of organisations to win the trust of people. It is in these numerous approaches of looking things that we can realize behavioural, organisational and situational factors that influence the use of GIS and its impact on sustainable development.

Role of Geospatial Information Systems in Sustainable Development

Environmental Monitoring and Conservation

GIS is a significant element of monitoring the environment since it allows scientists to accomplish high-resolution spatial analysis that enables them to make judgments regarding the protection of the environment, ecosystem control, and environmental legislation. Contemporary GIS systems keep a close watch on the vegetation cover changes, land degradation, biodiversity habitats, and shielded territories through remote sensing information, satellite images, and geospatial analytics. This assists the scientists and those in charge to identify new ecological issues at an early stage, such as habitat fragmentation, deforestation, erosion, and invasion of species. Remote sensing technologies augment this by providing continuous, multispectral, and time-series information to quantify ecological information including NDVI (Normalised Difference Vegetation Index), carbon stock distribution and forest canopy health (Coops & Wulder, 2019). These lessons will be important in coming up with targeted conservation initiatives, measurements of adherence to natural regulations, and increased sustainability of forests and watersheds. GIS also contributes to the efforts aimed at restoring ecosystems with the help of identifying the areas of the damaged ecosystems, setting priorities in the restoration areas and monitoring the ecosystems recovery in the long-term.

Climate Change Mitigation and Adaptation

GIS can now be quite useful in the study and planning of climate change as it can be checked on the threats, weaknesses, and threats of climate change at very numerous angles. GIS can be combined with climate models to allow the scientist to observe the alteration in temperature, precipitation, and extreme weather events when the climate changes. This spatial intelligence assists in finding locations where heat waves, floods, drought, and coastal flooding have a high probability of occurring (Santamaria et al., 2020). The geospatial vulnerability assessments assist governments in identifying gaps in infrastructure and economy, which plays a role in ensuring that the climate adaptation plans are targeted to address the communities that are in the highest risk. The use of GIS-based scenario modelling is increasingly having a role in policymakers wishing to understand the impact of climate activity, such as altering more green space, rehabilitating watersheds, or making housing policies more resilient. GIS assists individuals to make improved choices, increases the precision of the adaptation planning and fosters the formation of national climate action schemes, which adhere to the worldwide agreements, because it shows how hazards of climatic character change with both space and time.

Sustainable Urban Development

GIS is a transformer to make cities more sustainable and resilient as it enables us to utilize facts in enhancing urban planning and infrastructure. GIS assists planners to determine whether the land is suitable to some purposes, approximate the expansion of cities, and manage conflicting desires on space in traffic areas using spatial modelling (Yeh et al., 2017). It enhances planning on transportation by identifying the best routes to use, considering traffic trends and design of the public transport systems that reduce pollution and traffic congestions. GIS can also be used to build infrastructure that consumes less energy by identifying the best places to build green buildings, renewable energy facilities and land-use identifications that can withstand even climate change. GIS is used to help in the housing policy by examining the ease of access to parks, government services and other social amenities. This ensures that development is in compliance with the provisions of fairness and sustainability. Air quality monitoring systems that integrate with GIS are also capable of assisting cities in locating areas that have a high level of pollution. This will allow them to act to enforce the law and make people listen. Overall, GIS is encouraging a smart city concept since it assists cities to become greener in a manner that considers their geography.

Renewable Energy Optimization

GIS has, to a great extent, assisted in planning projects based on renewable energy. It also examines geographic, environmental, and technical factors in order to find the best place to place solar, wind, hydro, and biomass energy installations.

Multicriteria decision analysis (MCDA) which is GIS-based considers factors such as land availability, terrain, solar radiation, and wind speed, distance to transmission networks, and environmental restrictions to ensure energy sites are long-lasting (Ali and Mohammed, 2021). This plan is cost effective in the sense that it targets areas with infrastructures and avoids sensitive places to protect the environment. Moreover, GIS-based modelling assists individuals to make more intelligent investment decisions by predicting the amount of power that a project would produce, the future performance level, and its implications to the environment. With the shift to less-carbon economies in countries, GIS would be an excellent method to design how to utilize energy in an area, utilize clean energy in intelligent fashions, and discover how to develop without damaging the natural environment.

Risk Reduction

GIS plays a significant role in disaster risk reduction (DRR) as you would be able to view the dangers, exposure, and vulnerabilities in space prior, during and after a disaster. GIS allows the formation of early warning systems by uniting real-time data on satellites, sensors and the field that allows predicting floods, landslides, wildfires, earthquakes, and cyclones (Pham et al., 2022). The hazard simulation models may aid the disaster management agencies in identifying locations that are likely to be hazardous, establishing evacuation routes, storing emergency supplies, and developing back-up plans. Following a disaster, GIS can be used to determine the extent of the destruction, build infrastructure, and organize humanitarian aid by mapping the affected people, lost assets, and challenges with the supply chain. With GIS, communities can be empowered to be more robust, ready and can implement evidence-based and globally compatible disaster risk reduction (DRR) policies.

Agriculture and Food Security

GIS assists the farmers to choose better as it allows farmers to control the crops, soil, water and other farm inputs more precisely and spatially. Precision agriculture: It involves GIS and remote sensing to have an eye on crop development, identify factors that cause stress to crops, including pests or water scarcity, and observe how the soil is being altered in nutrition (Li et al., 2020). This assists farmers to maximize the use of their water, fertilizer and land management that reduces wastage and raise their output. GIS can also be used in food security planning to monitor the change in agricultural outputs, identify the locations susceptible to climate change, determine the amount of food to be collected, and analyze the effect of drought and floods to food supply chain. GIS improves national strategies to make food resilience and sustainable agricultural growth better by bringing together climatic data and agricultural models.

Governance and Public Participation

GIS has increasingly gained popularity as a strong instrument of enhancing governance, transparency and citizen involvement in sustainability programs. Participatory GIS also allows individuals within a group to map the issues within their area, trace developments within their environment, and provide planners with what they know. This renders geospatial decision-making a more democratic process (Brown and Kytta, 2018). The strategy is most effective when the marginalised individuals or indigenous communities possess valuable information on how to use land, how to manage the resources or how to solve the problems of the society. Individuals are also more responsible when GIS is adopted in governance since it provides them spatial evidence that makes them make their decisions on such issues as public policy, budgetary allocations, environmental requirements, and service delivery. Participatory GIS promotes people collaborating to make choices, empowers a broad group of stakeholders and assists in sustainable planning which is fairer and socially accountable.

DISCUSSION

The findings indicate the relevance of GIS in ensuring sustainable development that is evidence-based. They also demonstrate that there is a significant variation in the use of the geospatial tools both across industries and geographies. In developed economies, GIS is being extensively applied in climatic modelling, city systems, and the environmental monitoring. However, in many poor nations, the issues of obtaining the necessary data, among others, and locating people with the right set of skills, as well as strengthening their organizations, are more significant. These structural deficiencies complicate the GIS to contribute to equitable results to sustainability. The paper highlights the necessity of interdisciplinary solutions due to the general fact that the problems of sustainability rarely reside on their own; instead, they might need integrated geospatial frameworks, which combine environmental, economical and socio-cultural aspects in coherent decision support systems.

An important development that emerges through the literature is that GIS is no longer simply an instrument of mapping. It is also being used as real time, predictive analytics and information simulations currently. GIS, used together with AI, cloud computing, and big data solutions, can turn into a mighty instrument of predicting climate risks, utilizing the land and resources as effectively as never before, and modeling the future more precisely than ever before. However, using such types of intricate systems complicates life of the businesses, including the necessity to find new methods of doing things, software to increase capacity, and funds to invest in digital infrastructure. These issues are the manifestations of larger issues that have been discovered in research on technology integration, including the fundamental article that discusses the significance of flexibility and adaptability of organisations and their leadership to facilitate the work of technological changes.

The other consideration is that there are moral and governance issues that arise when a large amount of geographical information is employed. With increased GIS relying on spatial imagery of higher eventuality, mobile location-tracking, and sensors based on the Internet of Things, individuals are increasingly concerned about privacy, authorization, fairness, and the likelihood of geospatial monitoring. Unless the right norms and laws exist, geospatial technologies may increase power inequalities, neglect the vulnerable individuals, or result in non-proper utilization of the personal geographical information. Findings indicate that there are clear data governance, stakeholder engagement, and regulatory oversight among others, which are important to ensure that GIS is a way to ensure sustainability in a socially responsible and morally good manner.

Lastly, the discussion states the importance of community-based and participatory GIS tools in increasing the accessibility of sustainability planning to all people. Geospatial tools are available to everyone because of participatory GIS and volunteered geographic information (VGI). This allows individuals to share whatever they know in the area, monitor the environment as well as have a say in the utilization of space. Such types of approaches make them more real, spatial databases more accurate, and can involve people to collaborate on sustainability projects. To make GIS effective, organisations should consider the combination of the new technology with the social inclusion, organisation preparation and long-term capacity building. This combined approach will make geospatial technologies to deliver more than information; it is the empowerment of resilience, equitable development and in line with the overall global sustainability targets.

These findings indicate that GIS is increasingly becoming an important concept on sustainable development grounded on factual basis. Nonetheless, they also show how geospatial tools are used in rather diverse ways by different areas and sectors. In developed economies, GIS is being extensively applied in climatic modelling, city systems, and the environmental monitoring. However, in many poor nations, the issues of obtaining the necessary data, among others, and locating people with the right set of skills, as well as strengthening their organizations, are more significant. These structural deficiencies complicate the GIS to contribute to equitable results to sustainability. The paper highlights the necessity of interdisciplinary solutions due to the general fact that the problems of sustainability rarely reside on their own;

instead, they might need integrated geospatial frameworks, which combine environmental, economic and socio-cultural aspects in coherent decision support systems.

An important development that emerges through the literature is that GIS is no longer simply an instrument of mapping. It is also being used as real time, predictive analytics and information simulations currently. GIS, used together with AI, cloud computing, and big data solutions, can turn into a mighty instrument of predicting climate risks, utilizing the land and resources as effectively as never before, and modeling the future more precisely than ever before. However, by using such complex system businesses have a harder time like new workflow systems, capacity growth programs and a budget to allocate which can be used on digital infrastructure. These issues are the manifestations of greater issues that have been witnessed when conducting research on technology integration. The core work emphasizes on the role of agility in organisations and adjustment of leadership as a matter of successful technological change.

The other critical issue that needs to be considered is the ethical and administrative issues that emerge with the usage of a large number of geographical information. With increased GIS relying on spatial imagery of higher eventuality, mobile location-tracking, and sensors based on the Internet of Things, individuals are increasingly concerned about privacy, authorization, fairness, and the likelihood of geospatial monitoring. Without the appropriate norms and laws, geospatial technologies may increase the power differences between people or marginalize vulnerable groups or can be used inappropriately to utilize personal geographical information in one way or another. These outcomes have demonstrated the value of having clear data governance, participation of stakeholders and regulatory oversight to ensure that GIS is used to foster sustainability in a manner that can be socially acceptable and ethically right.

Lastly, the discussion brings out the importance of the participatory and community-based GIS tools to ensure that planning towards sustainability becomes accessible to all. Geospatial tools are available to everyone because of participatory GIS and volunteered geographic information (VGI). This allows individuals to share whatever they know in the area, monitor the environment as well as have a say in the utilization of space. These types of techniques make it more real, spatial information more accurate and people cooperate in creating better environmental projects. To ensure that GIS is functional as it ought to be, businesses must employ new technology in addition to social inclusions, preparation and long-term capacity building. This combination approach will ensure that geospatial technologies do not just provide us with information. They are also able to make us stronger, principled development, and compatible with larger global objectives regarding sustainability.

MANAGERIAL IMPLICATIONS

The management implications of the study have put forward the need by organisations, government and development agencies to restructure their planning systems, operational systems and governance systems so as to be in a position to realise the revolutionary potential of GIS in sustainable development. Managers should understand that geospatial technologies are not merely technical tools; they are strategic resources, which can affect the decision making process in various domains of society such as the provision of the public service, resource distribution, building new infrastructure, and management of the environment. To enable firms to leverage these skills, it requires a lot of expenditure on constant training on GIS-based decision support, remote sensing, and spatial analytics so that their employees can apply the talents fully. Your approach to your talent needs to change should you seek to locate and retain those talents skilled in the geospatial science, data governance and digital sustainability. Further, firms must integrate a holistic and cross-functional strategy where environmental surgery, and planning, IT department, and even community involvement offices collaborate without any issues. With the rise of the use of higher-resolution photos, mobile geolocation information, and AI-related analytics, the managers should establish firm principles of how they deal with privacy, data quality, and ethical utilization of the information. To maintain trust and protect vulnerable communities and prevent the exploitation of sensitive spatial information, there must be clear guidelines on how to obtain, store and gain access to data. In addition, the management is advised to employ participatory and inclusive practices such as community mapping, citizen generated data as well as stakeholder consultations to make the process more legitimate and everyone willing to contribute towards the outcome of sustainability. It implies that GIS is not merely a technical system, but a mechanism that make government more transparent, people more powerful and come up with good decisions that benefit everyone.

THEORETICAL IMPLICATIONS

The theoretical implications complement the existing academic perception of the concepts of sustainability, integration of technology, and space decision-making by setting GIS as the cornerstone of the investigation and administration of intricate socio-environmental frameworks. Results in this paper solidify the systems theory by showing how GIS realises the interdependence of the environmental, social and economic aspects of sustainability through a more holistic perspective that cannot be dramatically captured in non-spatial models. The paper also brings developments to the existing socio-technical systems theory by integrating GIS with artificial intelligence, big data analytics, and cloud computing, thus demonstrating how technological artefacts, institutional frameworks, human behaviour, and governing mechanisms interact. This broadens the thinking on operational change and organisational adaptation, in that, spatial intelligence has become a significant subject matter in talent in modern organisations facing sustainability issues. The study contributes to the geographic decision-support theory because it shows the importance of real time analytics, predictive modelling and scenario simulation in decision-making situations. It is also a contribution to the literature on participatory governance, theorising on how community-based GIS and volunteered geographic information could result in production of knowledge that is more democratic, results in change of power and in improved long-term outcomes. Lastly, theoretical implications provide an indication of the necessity to introduce a new concept of the design that comprises equitable design, inclusive thinking, and responsible geospatial data practice. All these contributions are combined in highlighting the growing significance of GIS as a technological and theoretical foundational cornerstone of understanding and accomplishment of sustainable development in a fast-evolving global environment.

SCOPE FOR FUTURE WORK

Prospective studies of the application of GIS in sustainable development offer great possibility to develop improved theoretical knowledge, methodological advancement, and expand practical implementation in the fields of environmental, social and economic aspects. One of the most important opportunities of future studies consists of the combination of artificial intelligence, machine learning, and deep learning with geospatial analytics to develop intelligent self-learned GIS models that should provide real-time monitoring and self-assessments of sustainability. Such changes may end up causing the early-warning systems to act dissimilarly when it comes to extreme climatic conditions, biodiversity loss, and exposing people to health risks. There is also increased interest among people about the process of mixing geospatial big data of satellite constellations, IoT sensors, unmanned aerial vehicles, and citizen information to generate accurate, interoperable, and ethically managed results. With comparable geospatial data standards, it would be far simpler that the governments can collaborate to address global issues such as global warming, transboundary water management, and ensuring that everyone has access to sufficient food.

Future directions should concentrate on the development of equity-based and justice-oriented GIS frameworks that specifically aimed at answering the question of disproportionate environmental burdens of disadvantaged and vulnerable people. It will involve conducting additional research on participatory GIS, community mapping and digital inclusion practices to ensure that the geospatial technologies are beneficial and not detrimental to those who are already less privileged. Another interesting way of going is to make digital twins of cities, ecosystems, and infrastructure networks using capitalism of GIS. This would enable high-level scenario modelling, prediction of sustainability and the planning of climate resilience. Such virtual environments may provide the decision makers with useful tools to observe the future of urban policies, land use, or conservation initiatives. Moreover, the future studies are to examine the structure of governance and institutional or institutional configurations that foster successful implementation of GIS with emphasis on capacity building, alignment of policies, and inter-agency collaboration. It is especially important to places with the low incomes where it is difficult to apply GIS as people are not aware of how to use it and infrastructure is not good.

The other important area on which future researches should focus is the ethical principles of using geographical data on exploration. Now more than ever, it is more crucial to consider the problem of privacy, artificial intelligence-solutions related to surveillance, data ownership, and algorithm ology bias in GIS ecosystems where high-resolution spatial data and artificial intelligence-enhanced geolocation analytics now make it everywhere. Professionals are advised to explore ways of enhancing transparency, honesty and trust in geospatial systems and ensure that the process of sustainability would not unwillingly undermine civil liberties. Furthermore, further studies are necessary to extend the use of GIS to new spheres of sustainability, i.e., the circular economy, the blue economy, the maximization of renewable energy, and the climate financing. Combining geospatial intelligence with macroeconomic indicators, behavioural measurements and ecological measurements, researchers are able to develop more comprehensive models of tracking progress towards the Sustainable Development Goals (SDGs). Lastly, people in other areas, including geospatial science, environmental psychology, public policy, engineering, and social innovation, will be required to make new GIS tools and conceptual models. The reason is that the world is getting more global, and climate change is becoming sensitive.

CONCLUSION

The Geospatial Information Systems have proven to be fundamental in facilitating sustainable development, whereby it offers unequal capabilities of gathering, analyzing and revealing spatial distributions of the environmental, social and economic systems. With the sustainability issues becoming more difficult to manage due to climate change, it means that urban expansions, resource loss and social disparities, GIS can provide the spatial data that can enable policy makers to come up with policies that are intelligent, robust, and future ready. GIS with AI, big data analytics, remote sensing and digital platforms have rendered it even stronger. This now allows it to perform tasks such as real time monitoring, predictive modelling, and multi scenario simulation that transforms the way governments, businesses and communities perceive and respond to key issues to do with sustainability. The combination of various datasets and participation through enhanced evidence-based decision-making and transparency makes GIS instrumental in fostering sustainability transitions globally, which explains why it can be regarded as a key source of catalyzing sustainability shifts. We must however invest smartly in institutional capacity, ethical government to remain human, technology infrastructure and interdisciplinary collaboration to achieve the best out of GIS. You must not only be good with technology to make good use of it, but the systems should also be in place to assure protection of privacy, equal access to everyone, and social capital in geospatial systems. To keep GIS in operation to achieve a sustainable development, large scale, organisations must develop cultures that embrace innovation, cross sector operation and long-term stability. With the ever-improving geospatial technologies, they can do a lot in transforming the future of how we plan our sustainability, foster adaptive management, and assist societies in managing environmental uncertainties more time and cost-effectively and proactively. After all, GIS is more a potent instrument of change and an intelligent means of reaching a more responsible, more resource-efficient, and environmentally safer future.

REFERENCES

Ali, A., & Mohammed, S. (2021). Using GIS-based multicriteria analysis for renewable energy site selection: A review. *Renewable & Sustainable Energy Reviews*, *143*, 110–120.

Brown, G., & Kyttä, M. (2018). Key issues and research priorities for public participation GIS (PPGIS): A synthesis based on empirical research. *Applied Geography (Sevenoaks, England)*, *95*, 123–136.

Coops, N. C., & Wulder, M. A. (2019). Breaking the habit: Remote sensing in ecosystem monitoring. *Ecological Indicators*, *98*, 1–5.

Fang, Y., Nie, Y., & Penny, M. (2021). Geographical modeling of COVID-19 transmission using GIS and spatial statistics. *International Journal of Health Geographics*, *20*(1), 1–12. PMID: 33397338

Goodchild, M. F. (2018). Reimagining the history of GIS. *International Journal of Geographical Information Science*, *32*(12), 1–20.

Hidayati, D.. (2020). Slum mapping using remote sensing and GIS: Supporting urban resilience. *Cities (London, England)*, *96*, 102–115.

Kabisch, N.. (2021). Mapping urban nature for sustainable planning: Advances in geospatial methods. *Landscape and Urban Planning*, *214*, 104–136.

Kumar, L.. (2014). Land degradation mapping using object-based classification in GIS. *Environmental Monitoring and Assessment*, *186*, 8451–8469.

Li, Z., Chen, Y., & Li, B. (2020). GIS-based precision agriculture: A review of methods and applications. *Computers and Electronics in Agriculture*, *173*, 105–128.

Longley, P. A., Goodchild, M. F., Maguire, D. J., & Rhind, D. (2011). *Geographic information systems and science* (3rd ed.). Wiley.

Malczewski, J. (2010). Local spatial decision support systems. *International Journal of Geographical Information Science*, *24*(5), 1–25.

Meadows, D. (2011). *Thinking in systems: A primer*. Chelsea Green.

Musa, S.. (2015). GIS-based poverty mapping: A global review. *International Journal of Geographical Information Science*, *29*(10), 1–19.

Pham, B. T.. (2022). GIS-based modeling for natural hazard susceptibility using machine learning. *Geomatics, Natural Hazards & Risk*, *13*(1), 345–366.

Santamaria, M., et al. (2020). GIS-based climate change modeling for resilience assessment. *Environmental Research Letters, 15*(9), 094–112.

Venkatesh, V., Thong, J. Y. L., & Xu, X. (2012). Consumer acceptance and use of information technology. *Management Information Systems Quarterly*, *36*(1), 157–178.

Vinuesa, R.. (2020). The role of artificial intelligence in achieving the Sustainable Development Goals. *Nature Communications*, *11*, 1–10. PMID: 31932590

Yeh, A. G. O.. (2017). GIS-based modeling for urban sustainability and smart growth. *Environment & Planning B*, *44*(6), 1012–1035.

Zhao, F., Li, X., & Xu, X. (2023). Integration of AI and GIS for environmental prediction. *ISPRS Journal of Photogrammetry and Remote Sensing*, *197*, 45–60.

Chapter 8

Geospatial Information Systems-Enabled Logistics for Sustainable Supply Chain Management

Hongmei Chen
School of Economics and Management, Yanshan University, Qinhuangdao, China

Jahangeer Ahmed Soomro
https://orcid.org/0000-0003-3168-3265
School of Economics and Management, Yanshan University, Qinhuangdao, China

ABSTRACT

Geospatial Information Systems (GIS) are increasingly serving as a fundamental facilitator of sustainable logistics by prioritizing spatial considerations in supply chain decision making. The integration of GIS with optimization techniques, remote sensing, telematics and artificial intelligence enables organizations and municipalities to enhance routing and facility placement, spatially analyze environmental and social effects, and bolster resilience against climate and infrastructure disturbances. Empirical studies indicate that GIS-based routing and dispatch systems can significantly decrease vehicle kilometers traveled, turnaround times, and carbon emissions in last-mile operations, especially when integrated with multi-objective optimization that incorporates emissions and exposure metrics. Simultaneously, GIS facilitates environmental management and energy infrastructure planning

DOI: 10.4018/979-8-3373-6320-2.ch008

by superimposing ecological, demographic, and infrastructural layers to mitigate exposure and land-use effects. Advancements in AI and telemetry data enhance the temporal granularity of decision making.

1. INTRODUCTION

Geospatial information systems (GIS) have now transcended the traditional mapping and visualization tools into highly integrative and flowing decision support systems which have radically transformed the design, operation and assessment of logistics and supply chains. Modern day supply chains are faced with mounting pressures due to increased e-commerce demand, urbanization, global warming, limited resources, competition and environmental and social responsibility demands of the stakeholders. These pressures require spatially conscious and data driven solutions (Bui et al., 2021; Wang & Hsu, 2024). GIS is the only tool that enables the combination of locational data with both the time, environmental, infrastructural and socio-economical data. This will inform route optimization, facility location, demand prediction, resilience planning and environmental footprint. These features make GIS specially desirable to support the sustainable supply chain management goals, such as emissions reduction, resource efficiency, reducing land-use effects and improving goods and services availability (Hammitzsch & Lendholt, 2011; Nimmagadda et al. 2019).

The fact that GIS has been integrated into logistics is not just a technical enhancement, but it is a shift in paradigm into place-based decision making within the supply chains. The traditional supply-chain models often use aggregate demand and cost curves that ignore the effects of spatial heterogeneity and network-topology (Patel & Desai, 2019). Conversely, GIS explicitly represents geographic constraints, including road networks, land-use and the distribution of suppliers and consumers, along with the sensitivities of the geographies, making them optimize and scenario-analyze to consider spatial trade-offs (Yokota & Pagkalinawan, 2022). The use of this spatially explicit lens is important when sustainability effects are in focus, since both environmental externalities (CO^2 emissions, air-pollution exposure and habitat fragmentation) and social-equity externalities (access to essentials, distributional burdens) are spatial in nature (Esfahbodi et al. 2023). Concisely incorporating geospatial analysis into logistics transforms the unit of analysis of aggregated flows to place-sensitive flows, thus enhancing the performance of operations and sustainability analysis.

It is possible to distinguish three general GIS-enabled capabilities that are of particular interest in sustainable logistics, such as spatial optimization (routing and facility location), mapping and environmental impact and risk assessment and resilience (Sowmiya & Manimaran, 2024; Weng et al., 2024). First, spatial optimization utilizes the network models and geospatial data to reduce travel distances, time, energy consumption, or emissions and meet the service constraints. Empirical research shows that route planning using GIS and vehicle routing integration decrease the total distance and emissions, particularly in intricate urban last-mile settings where traditional heuristics are complicated by congestion and access controls, as well as dynamic demand (Lehoullier, 2023; Plazier et al., 2024). In addition, integrating GIS with other sophisticated algorithms, such as genetic algorithms, mixed-integer programming and machine learning, will allow more realistic and multi-objective optimization, integrating environmental indicators with cost and service standards (Qu, 2024). The green supply-chain practices on the operational scale are based on the principle of such multi-criteria spatial optimizations (Yachai et al. 2021).

Second, GIS offers powerful techniques of environmental-impact mapping and lifecycle spatialization. Conventional lifecycle assessment and carbon accounting tend to hide the origin of emissions space-wise and how they interact with vulnerable human groups or sensitive ecosystems. Geospatial techniques enable scholars and practitioners to distribute emissions and environmental harms over the landscape and population, recognise hotspots (freight corridors with a high level of pollution) and implement interventions (consolidation centers or low-emission zones), which target those hotspots (Scott & Rajabifard, 2017; Kochanek et al., 2025). Moreover, incorporating remote sensing and land-cover information and socio-demographic strata, GIS enables the holistic evaluation of the environment that connects logistics decisions with biodiversity, water, and land-use externalities, and thus adapts supply-chain planning to the wider sustainability models (Susanty et al. 2016; Nimmagadda et al., 2019).

Third, GIS increases supply-chain resilience in the form of spatial risk analysis and scenario modeling. Extreme events (floods, storms, heat waves) and climate change have a material impact on transportation infrastructure and nodal facilities. Geospatial analysis makes possible vulnerability mapping (intersection of hazard maps with supply-chain assets), single points of failure finding in networks and stress testing of alternative routing and inventory options in the face of spatially explicit disruption (Gopal et al., 2022; Rezvani et al. 2023). The current evidence indicates that integrating GIS with AI and optimization enhances the ability of supply chains to restructure in case of disruption and achieve sustainability goals, such as minimizing detours that would otherwise shoot up the emissions or resource consumption (Zaroujtaghi et al., 2025; Agoo et al., 2024). In this regard, GIS will be an efficiency as well as an adaptive, sustainable logistics planning tool.

In addition to technical roles, the institutional and organizational aspects of the GIS adoption are important towards sustainable logistics. A lack of technical capacity, broken data governance, proprietary data silos, and a mismatch between short-term commercial objectives and long-term sustainability interests are some of the barriers to adoption (Tsakiridi, 2021). However, where companies and cities invest in spatial data infrastructure and cross-sectoral data exchange, there are returns of a substantive nature: better facility-location applications that reduce lifecycle emissions, better urban consolidation plans that reduce the number of last-mile vehicle kilometers covered and fairer service provisioning plans that eliminate so-called logistics deserts. The literature underlines the use of GIS in the organizational decision processes, such as ascertaining GIS-based scenarios and the procurement, network design, and sustainability reporting systems (Deng et al. 2019).

The modern research edge of urban and regional governance deals with the operationalization of GIS in various areas of stakeholders. A significant source of urban freight emissions and congestion, last-mile delivery is at the interface of the activities of the private logistics operators, municipal planners and citizens (Ha et al. 2023). GIS offers a common analysis platform upon which to trade-off (service levels against emissions, access against traffic) and in which to implement collective policy tools, such as low-emission zones, micro-hubs and dynamic curb-space allocation (Qu, 2024). Both the potential and the complexity of this solution are emphasized by empirical studies on the city level: GIS analyses may be used to suggest the course of digressions or consolidations with clear sustainability advantages, but their effective execution requires regulatory support, coordination of stakeholders (Maxner et al., 2025; Plazier et al., 2024). These results emphasize the need to adopt socio-technical solutions that integrate geospatial analytics innovations into governance.

The GIS toolkit to sustainable logistics is being increased by methodological advancements. The integration of GIS with big data streams (live telematics, movement traces of mobile phones), remote-sensing data and machine-learning algorithms allow much more detailed modeling of the demand trends, traffic flows and environmental exposures than has been commoditized before (Tsakiridi, 2021). It is proven that congestion predictions based on high-frequency location data combined with GIS network models can be improved and dynamic routing is more efficient when environmental goals, such as population-weighted exposure to emissions are incorporated into objective functions (Hosen et al. 2023). However, the application of such data awakes the issue of privacy, algorithmic transparency and fair representation.

The chapter can be considered important since it clarifies the central importance of Geospatial Information Systems (GIS) in making logistics a more sustainable, information-driven and resilient element of the contemporary supply chain management. With increasing needs to reduce carbon, be regulatory compliant, resistant to disruption and reduce congestion in cities, GIS provides an inimitable spatial intelligence platform, which facilitates informed decision-making. GIS can be used by organizations to optimally plan routes, reduce transportation emissions, determine environmental hotspots, and place facilities strategically to mitigate ecological footprints by combining real-time geospatial data with environmental indicators, network topology and socio-economic variables. In addition, the chapter highlights the role of GIS in supply-chain transparency, risk evaluation, and climate-resilient logistics planning all of which are necessary to attain the long-term sustainability objectives. Combining the insights gained through peer-reviewed studies and other emerging technologies, this chapter provides scholars, practitioners, and policymakers with an overall idea of how GIS can be a strategic enabling force of sustainable logistics as well as a pillar that supports greener and more socially responsible supply-chain systems.

Figure 1. GIS for logistics and supply chain

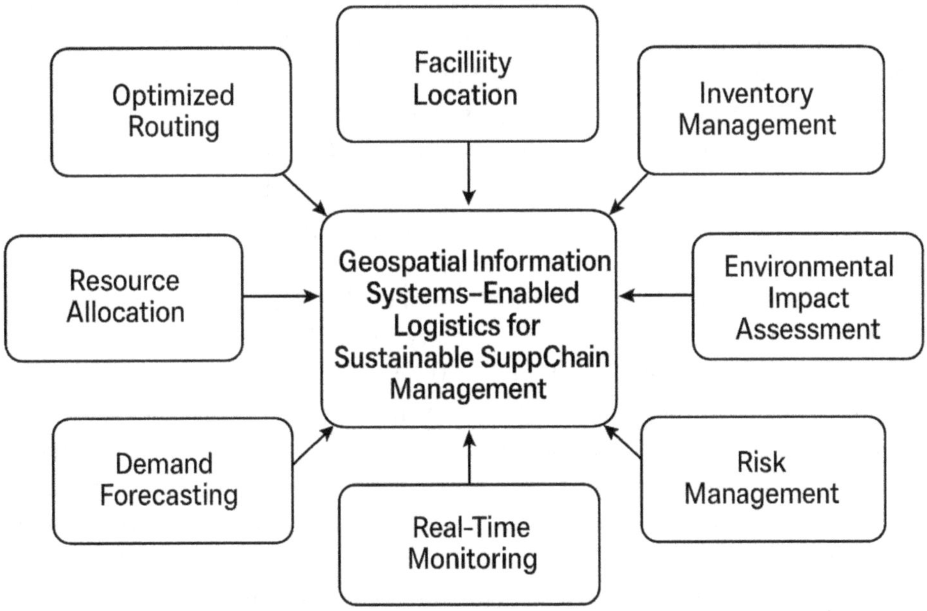

2. LITERATURE REVIEW

Geographic Information Systems (GIS) have now grown into essential facilitators of spatial conscious logistics and supply-chain decision-making. GIS supports location decisions, routing, demand allocation, risk assessment and environmental hot-spotting. These are critical in sustainable supply chain management (SSCM) by enabling the blending of spatial layers. These layers include transport networks, land use, environmental exposures and socio-economic data, with optimization models and real-time telemetry (Agoo et al. 2024). Recent reviews and explicated researches indicate that GIS can minimize the vehicle kilometers, specialize exposures and emissions, and enhance resilience against spatially explicit risks of flood and storms which align the operations performance with the environmental and social purposes.

The literature identifies a number of priority areas on future research and practice to maximize the application of GIS enabled logistics to sustainability. To make comparisons and regulatory reporting meaningful, the first thing that is needed is a standardized methodological framework that will allow environmental and social impacts to be spatialized along supply chains (Maurya & Kumar, 2024). Secondly more attention needs to be paid to the governance models that need to accompany GIS-informed interventions in the urban setting such as incentive systems and public-private collaborations (Falasco et al. 2024). Third, the fair aspects of GIS applications, that optimization is not overwhelmed by marginalized communities. It need to firmly incorporate socio-spatial metrics into optimization algorithms (Davis et al. 2021). Fourth, the local authorities and small logistics providers should also be provided with capacity building to spread the GIS benefits beyond the large multinational companies. The solution of these areas will convert the technical capability of GIS into quantifiable sustainability outputs in actual supply-chain systems (Malakar & Roy, 2024).

Overall, GIS-enabled logistics is a potential evidence-based avenue to redefine supply-chain activities in terms of sustainability goals. GIS operational optimization and environmental stewardship (Raihan, 2024). By making space a key concept of modelling, GIS mitigates unnecessary travel, locates emissions and hotspots of impact, supports resilience to spatially diverse risks and promotes collaborative administration structures needed to foster sustainable city freight (Malaker & Meng, 2024). Because scientific and practitioner communities are still working toward developing integrated geospatial, algorithmic and governance strategies, GIS can be further used as a cornerstone of sustainable supply chain management with limited resources and rising environmental pressure (Ekeh et al. 2025).

3. METHODOLOGY

This chapter adopts qualitative research method via discussion that how different sectors/ industries can use GIS (Geospatial Information Systems) to do best in their logistics that are energy efficient and route optimized. This discussion constructs to the sustainable supply chain via GIS enables logistics. The chapter also discuss challenges and opportunities that come with GIS enabled logistics and supply chain.

4. INDUSTRY WISE DISCUSSION

4.1 Agriculture

Precision agriculture and sustainable agri-logistics are made possible by a collaboration between GIS and remote sensing. GIS assists agriculture to map soil variability, crop health (through multispectral remote sensing) and irrigation requirements and reduce the waste of inputs and transport demand (fertilizer and pesticides) (Logeshwaran et al. 2024). GIS is also used to plan farm-to-market paths, isolating points, de-icing and cold-chain connections to perishable goods to reduce the loss of crops after harvest, as well as transportation emissions (Ghosh & Kumpatla, 2022). Recent systematic reviews demonstrate that field sensorization and decision-support based on GIS and UAV images result in improved yield prediction and resource procurement, whereas the implementation is not evenly spreading in the contexts of low-and-middle-income because of the gaps in data and infrastructure and capacity (Mathenge et al. 2022). The literature shows that spatial analytics minimize unnecessary journeys and enable efficient logistics through improving harvesting, storage location, and route arrangement (Yusianto & Hardjomidjojo, 2020).

4.2 Energy (Petroleum, Coal, Shale Gas)

Movement of fuels, maintenance of sites and maintenance of network in the energy industry is spatially occupied and environmental sensitive. Some of the areas of GIS use in energy logistics include fuel tanker routing, storage terminal locations, pipeline right-of-way and resulting accident or spill evacuation/response (Jeong et al. 2021). Due to the environmental layers, including wetlands and population centres, GIS allows planners to choose the transport alignments that will reduce the risk of ecological and human exposure, which will assist in the regulatory compliance and in the social sustainability (Kochanek et al. 2025). In the extractive industries like the shale gas, the GIS is used to optimize the location of well-pad to reduce land fracture and road constructions, and to simulate the traffic movement of the rigs to the processing plants to minimize the local air-quality and traffic effects (Korpinen et al. 2023). Recent developments utilize remote sensing in addition to persistent surveillance of fugitive emissions and infrastructure durability to achieve a more precise upkeep and reduce the number of unforeseen freight transfers, which results in a quantifiable form of sustainability payoff (Ahmada, 2024).

4.3 Manufacturing & Retail

GIS is used extensively in manufacturing and retail in the location of facilities, network design, decentralization of inventory and the last-mile logistics. The GIS spatial optimization models help businesses to choose the location of warehouses and distribution-centers, which have the least lifecycle transport emissions and meet the service-level constraints (Song & Ma, 2024). In retail, GIS aids store assortment, demand catchment analysis and route optimization of replenishment and e-commerce fulfilment. These spatial analyses also facilitate the consolidating strategies in the form of micro-hubs to minimize last-mile vehicle-kilometers travelled (VKT) and urban emissions (Pavani et al. 2024). It is demonstrated in the literature that multi-objective solutions can be achieved by integrating GIS of mixed-integer programming, algorithms and real-time traffic feeds, so managers can operationalize greener trade-offs as opposed to using cost-only metrics only (Formánek & Sokol, 2022).

4.4 Healthcare & Pharmaceuticals

The supply chains in healthcare are critical to life and spatial. Vaccine distribution, cold-chain planning, facility catchment analysis and emergency logistics have all been implemented with the use of GIS (Bridgelall, 2023). Recent literature shows that GIS application has been applied in optimization of distribution routes of temperature sensitive pharmaceuticals, locating cold storage nodes, mapping service accessibility of vulnerable groups (special needs, rural dwellers), increasing equity but minimizing unnecessary or emergency travel that enhances emissions and expenditures (Thakur-Weigold et al. 2022). Big-data analytics combined with GIS allows mapping of the demand (outbreaks, season spikes) almost in real time and route modifications to emergency medical deliveries (Muyingo et al. 2022) . Notably, the authors focus on regulations, cold-chain validation, as well as the privacy of data and accuracy of demand predictions as a limitation to complete GIS implementation in health logistics (Shamsuzzoha et al. 2020).

4.5 Construction

Construction projects are manifestations of intricate temporal and geographical logistics material deliveries, waste disposal, site access, focused within small areas of space and limited time. GIS with help of Building Information Modelling (BIM) and logistics planning becomes one of the ways to optimize delivery schedules, curbside utilization and material staging (Uzairuddin & Jaiswal, 2022; Akindele et al., 2025). This decreases queuing, re-deliveries and idle time, which in turn reduces emissions and enhances sustainability of the project. The latest suggestions on GIS-based BIM integration create digital twins of construction sites, allowing to simulate the route of the deliveries, access of the vehicles and on-site management to exclude distraction and numerous movements (Bansal, 2020). The literature also indicates significant reduction in emission and cost savings when spatial simulations are used to guide deliveries that are made in batches, as well as to lower the time on site; interoperability and real-time integration of data are technical challenges (Yu et al. 2021).

4.6 Defense & Military

GIS has a long history in military logistics since contingency routing and terrain analysis require high spatial intelligence, as well as strategic mobility. Military GIS studies in the present focus on the concept of geospatial intelligence (GEOINT) to supply chain routing in settings of contested or damaged infrastructure, autonomous convoy planning and quick re-routing of plans and predictive maintenance (Altarez et al. 2024). The ethical and legal aspects of the use of GIS are also discussed in the literature on defense (targeting versus humanitarian logistics), but in an approximate methodology, it follows how GIS is used to increase the resilience of logistics by detecting alternative routes, pre-positioning resources, and reducing the exposure of convoy to areas with high risks (Joos, 2022).

4.7 Wood Industry / Forestry

Forestry and wood products logistics built onto distributed landscapes that are characterised by variability of access, seasonal limitations and high sustainability demands (biodiversity, soil conservation). The literature on precision forestry and Forest 4.0 provides an idea of how GIS, remote sensing and IoT can be used to plan harvest, skid-road timings and transport paths. This minimizes soil disturbance and fuel usage (He & Turner, 2021; Damasevicius et al., 2024). GIS supports traceability and certification (chain of custody) via mapping of harvest origin, transport routes and mill destinations (Ahl et al. 2020). This improves market access to sustainably certified wood. Recent reviews say that digitalization, i.e. GIS integrated with blockchain and IoT can remove inappropriate backhauls, enhance load optimization which can help reduce emissions and enhance a better management of forest resources (Zhang et al. 2020).

4.8 Food Industry

Chains of food supply especially those involved in perishables have an acute sensitivity towards spatial and temporal variations. Geographic Information Systems

(GIS) are regularly used as methods of mapping cold-chain nodes, routing cold-chain transportation, urban food distribution networks and evaluating the access to food such as the mapping of food deserts (Odion et al, 2023; Changmai, 2023). Recent empirical studies have demonstrated that GIS-enabled consolidation centers combined with the dynamic routing algorithms can address the last-mile emissions as well as food loss. Also epidemiological and public-health nutrition research exploits GIS to map the areas of contamination hazards and the supply sources in events of outbreaks, hence, informing the quick response and reducing the wastage of precautionary recalls (Asgharizadeh et al. 2022). Systematic reviews also highlight that comprehensive spatial-temporal models that tie production variability to logistic and retail demand are needed that will reduce waste and emissions throughout the food value chain (Rejeb et al. 2021).

5. ASSOCIATED CHALLENGES

5.1 Data availability, quality and fragmentation

Such spatially explicit analytics require a high level of the resolution of infrastructural, telematics and socio-demographic data. These datasets are often however spread out in private companies, public departments and proprietary outlets thus making them to be expensive to purchase. The lack of data quality reduce the reliability of the models and limit its adoption especially by small or medium enterprises and towns with limited budgets. Lack of an interoperable standard of data further eats up integration costs (Musa et al., 2024; Delgado et al., 2025).

5.2 Technical capacity and skills gap

Expert capabilities required to do effective logistics with support of GIS include specialized GIS, network modelling, machine-learning/optimization and remote sensing. Lots of logistics companies, particularly smaller and third-party companies and local governments do not have in-house capability or access to trained staff, which limits uptake and promotes the over-dependence on out-source consultants (Ritchie et al. 2023). Such capacity limitations are known to be a major hampering periodically cited on the way to transformation of geospatial awareness into operational alteration (Agoo et al. 2024).

5.3 Institutional and governance barriers

The implementation of GIS-based suggestions, including consolidation centers, low-emission areas or re-routing plans, usually involves cross stakeholder coordination between private carriers, municipal planners and community organizations (Dixit et al. 2022). Poorly aligned incentives and lack of clarity in data-sharing are the primary reasons why the GIS-based solutions fail to transition beyond pilot projects and into permanent practice (Aladaileh et al. 2024).

5.4 Integration challenges with legacy systems and operational workflows

The logistics is often carried out based on the old Transportation Management Systems (TMS) and Warehouse Management Systems (WMS). The software interoperability and careful change management is required because the output of the GIS converts into real-time execution environments, driver processes, and contractor schedules to optimize routing, spatial risks heatmap features. Without a smooth integration, GIS suggestions either go under used or they are completely ignored (Chen et al. 2021).

5.5 Cost and ROI uncertainties

Initial capital expenditures on spatial data infrastructure, sensors, GIS licenses and training are high. To companies with small operating margins, the unreliable and sometimes lengthy payback periods create doubt among the decision-makers to invest without definite pilot proofs to show that it can produce visible sets of fuel consumption, time and emissions cut-backs (Sun & Fan, 2020). This reluctance is more so when it comes to low resource environments.

Figure 2. Related challenges

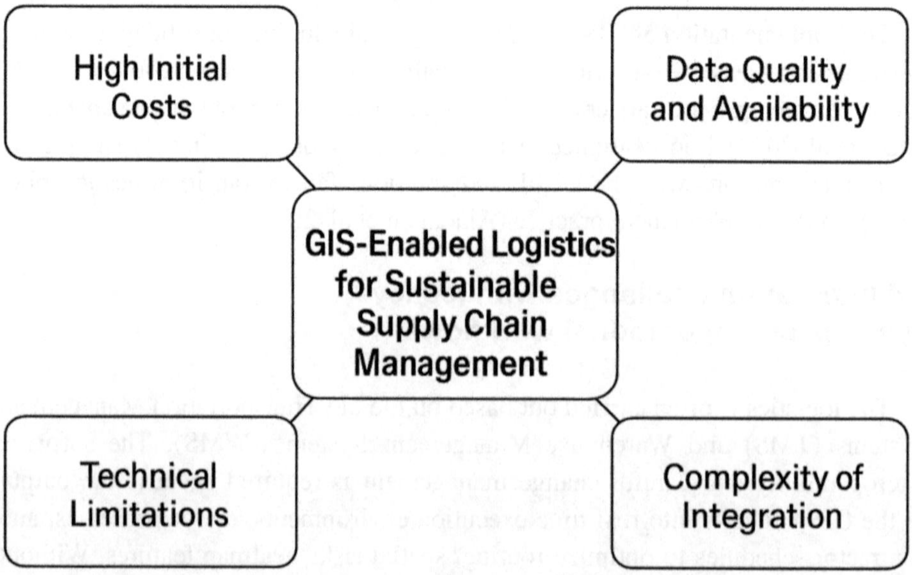

6. OPPORTUNITIES

6.1 Better environmental and social impact targeting

Supply planners can localize life-cycle impacts and exposure profiles, such as the location of freight-corridors through high-pollution areas, sensitive ecological locations or underserved communities (Chen et al. 2024). This specialization makes possible the intervention, in a specific manner and in a fair and equitable way, as with locating consolidation centers to reduce exposure in the most sensitive neighborhoods or by modifying the corridors to prevent ecological hotspots (Zailani et al. 2015). Compared with aggregated metrics, the defensible sustainability trade-offs produced by spatial hotspots are better.

6.2 Enhanced resilience and disruption management

Preplanning contingency, inventory pre-positioning, and fast re-routage in the case of a disruption are improved through the use of geospatial vulnerability mapping which is overlaying hazard layers on asset locations and network topology in order to reduce the cost of service failure and the cost of emissions penalty due to long rerouting caused by the disruption. GIS-AI hybrids speed up the process of recovery planning because it quickly identifies alternative corridors and staging points (Zhao et al. 2025).

6.3 Traceability, certification, and transparency

GIS supports the use of origin-to-consumer traceability in industries like forestry, food, and pharmaceuticals such that products that are sustainably certified have a greater market presence and that the supply chains promote the visibility of the chain. This spatial traceability enables the firms to meet market value needs of employees in terms of regulatory requirements and consumer expectations of provenance, thus creating a market value to greener logistics practices (Delgado et al. 2025).

7. CONCLUSION

The evidence-based case that GIS is not merely a mapping tool is provided by the peer-reviewed literature over the last two-three years. It is a cohesive platform that can make sustainable supply-chain management materially advanced in the case of integration with optimization, remote sensing, IoT and AI. The application of GIS-based systems in dispatch and route optimization, and spatial analysis in locating facilities and identifying hotspots to limit exposure, is associated with operational and environmental results that are measurable by sector-specific case studies: turnaround times, carbon footprints are decreased in pilot applications and spatial analysis enhances facility location and, respectively (Agoo et al., 2024; Bertolini et al., 2024). These results support the idea that integrating spatial intelligence into logistics allows making the trade-offs of the logistics, which include cost, service, emissions, equity, rather than the single-metric trade-off, which is one of the most crucial aspects when it comes to aligning commercial activities with climate and social objectives.

The role of GIS in resilience and risk management is also interesting. The spatial vulnerability mapping, that is, overlaying hazard projections to transport networks and critical nodes, allows planners to pre-position inventory, find the alternative routes as well as to stress-test networks in climate-related scenarios, thereby minimizing the time of service disruption and the penalty of emissions incurred during the long detours or emergency mobilization (Kochanek et al., 2025). Coupled convergence with AI and real-time telemetry will help increase responsiveness: dynamic routing and condition-driven maintenance reduce empty runs and encourage more efficient use of assets (Yadav et al. 2024).

In a bid to transform the technical potential of GIS to sustainable outcomes on the system level, three areas should be the main focus of research and practice. To begin with, the smaller carriers and municipalities will be included by means of the development and adoption of low-cost toolkits and interoperable standards of the spatial data. Second, the deployment of social-spatial indicators and equity controls underlying optimisation aims will protect unintentional redistribution of burdens. Third, by combining governance innovation, including incentive, regulation and common infrastructure with technical pilots, private incentive will be aligned with the long-term sustainability goals of the population. Provided that such priorities are taken care of, the literature suggests that GIS could be instrumental in minimizing logistics emissions, resilience, and more equitable and transparent supply chains.

REFERENCES

Agoo, J., Lanuza, R. J., Lee, J., Rivera, P. A., Velasco, N. O., Guillermo, M., & Fernando, A. (2024). Geographic information system-based framework for sustainable small and medium-sized enterprise logistics operations. *ISPRS International Journal of Geo-Information, 14*(1), 1.

Ahl, A., Goto, M., & Yarime, M. (2020). Smart technology applications in the woody biomass supply chain: Interview insights and potential in Japan. *Sustainability Science, 15*(5), 1531–1553.

Ahmada, I. (2024). Advanced sensors, monitoring. Smart Cities and Sustainable Manufacturing: Innovations for a Greener Future, 47.

Akindele, O., Ajayi, S., Oyegoke, A. S., Alaka, H. A., & Omotayo, T. (2025). Application of Geographic Information System (GIS) in construction: A systematic review. *Smart and Sustainable Built Environment, 14*(1), 210–236.

Aladaileh, M. J., Lahuerta-Otero, E., & Aladayleh, K. J. (2024). Mapping sustainable supply chain innovation: A comprehensive bibliometric analysis. *Heliyon, 10*(7). PMID: 38623205

Altarez, R. D. D., Apan, A., & Maraseni, T. (2024). The perspectives of remote sensing and GIS on military environmental impacts: A systematic review. *Environmental Monitoring and Assessment, 197*(1), 113. PMID: 39738966

Asgharizadeh, E., Jooybar, S., Mahdiraji, H. A., & Garza-Reyes, J. A. (2022). A novel travel time estimation model for modeling a green time-dependent vehicle routing problem in food supply chain. *Sustainability, 14*(14), 8633.

Bansal, V. K. (2020). Use of GIS to consider spatial aspects in construction planning process. *International Journal of Construction Management, 20*(3), 207–222.

Bertolini, M., De Matteis, G., & Nava, A. (2024). Sustainable last-mile logistics in economics studies: A systematic literature review. *Sustainability, 16*(3), 1205.

Bridgelall, R. (2023). Unlocking drone potential in the pharma supply chain: A hybrid machine learning and gis approach. *Standards, 3*(3), 283–296.

Bui, T. D., Tsai, F. M., Tseng, M. L., Tan, R. R., Yu, K. D. S., & Lim, M. K. (2021). Sustainable supply chain management towards disruption and organizational ambidexterity: A data driven analysis. *Sustainable Production and Consumption, 26*, 373–410. PMID: 33015266

Changmai, S. (2023). Geospatial Application for Dairy Supply Chain Management. *Journal of Genomics*, *17*(2), 174–183.

Chen, E., Zhou, Z., Li, R., Chang, Z., & Shi, J. (2024). The multi-fleet delivery problem combined with trucks, tricycles, and drones for last-mile logistics efficiency requirements under multiple budget constraints. *Transportation Research Part E, Logistics and Transportation Review*, *187*, 103573.

Chen, Y., Huang, Z., Ai, H., Guo, X., & Luo, F. (2021). The impact of GIS/GPS network information systems on the logistics distribution cost of tobacco enterprises. *Transportation Research Part E, Logistics and Transportation Review*, *149*, 102299.

Davis, J., Pijawka, K. D., Wentz, E., Hale, M., & King, D. A. (2021). Evaluating geodesign for community-based tribal planning: The role of planners in marginalized communities. *Journal of the American Planning Association*, *87*(4), 527–541.

Delgado, F., Garrido, S., & Bezerra, B. S. (2025). Barriers to Visibility in Supply Chains: Challenges and Opportunities of Artificial Intelligence Driven by Industry 4.0 Technologies. *Sustainability*, *17*(7), 2998.

Deng, Y., Gan, V. J., Das, M., Cheng, J. C., & Anumba, C. (2019). Integrating 4D BIM and GIS for construction supply chain management. *Journal of Construction Engineering and Management*, *145*(4), 04019016.

Dixit, A., Routroy, S., & Dubey, S. K. (2022). Analyzing the operational barriers of government-supported healthcare supply chain. *International Journal of Productivity and Performance Management*, *71*(8), 3766–3791.

Ekeh, A. H., Apeh, C. E., Odionu, C. S., & Austin-Gabriel, B. (2025). Leveraging machine learning for environmental policy innovation: Advances in Data Analytics to address urban and ecological challenges. *Gulf Journal of Advance Business Research*, *3*(2), 456–482.

Esfahbodi, A., Zhang, Y., Liu, Y., & Geng, D. (2023). The fallacy of profitable green supply chains: The role of green information systems (GIS) in attenuating the sustainability trade-offs. *International Journal of Production Economics*, *255*, 108703.

Falasco, S., Caputo, P., & Garrone, P. (2024). Can Geographical Indications promote environmental sustainability in food supply chains? Insights from a systematic literature review. *Journal of Cleaner Production*, *444*, 141100.

Formánek, T., & Sokol, O. (2022). Location effects: Geo-spatial and socio-demographic determinants of sales dynamics in brick-and-mortar retail stores. *Journal of Retailing and Consumer Services*, *66*, 102902.

Ghosh, P., & Kumpatla, S. P. (2022). GIS applications in agriculture. Geographic information systems and applications in coastal studies, 89-114.

Gopal, S., Staufer-Steinnocher, P., Xu, Y., & Pitts, J. (2022). Semiconductor supply chain: a 360-degree view of supply chain risk and network resilience based on GIS and AI. In *Supply chain resilience: Insights from theory and practice* (pp. 303–313). Springer International Publishing.

Ha, N. T., Akbari, M., & Au, B. (2023). Last mile delivery in logistics and supply chain management: A bibliometric analysis and future directions. *Benchmarking*, *30*(4), 1137–1170.

Hammitzsch, M., & Lendholt, M. (2011). Evolving an information logistics database for geospatial early warning systems. *Geomatics, Natural Hazards & Risk*, *2*(2), 95–109.

He, Z., & Turner, P. (2021). A systematic review on technologies and industry 4.0 in the forest supply chain: A framework identifying challenges and opportunities. *Logistics (Basel)*, *5*(4), 88.

Hosen, B., Rahaman, M., Kumar, S., Sagar, L., & Akhtar, M. N. (2023). Leveraging artificial intelligence and big data for advanced spatial analytics and decision support systems in geography. *Malaysian Applied Geography*, *1*(2), 62–67.

Jeong, H., Karim, R. A., Sieverding, H. L., & Stone, J. J. (2021). An application of GIS-linked biofuel supply chain optimization model for various transportation network scenarios in Northern Great Plains (NGP), USA. *BioEnergy Research*, *14*(2), 612–622.

Joos, G. (2022). Geographic Information Systems in Defense. In *Springer Handbook of Geographic Information* (pp. 685–705). Springer International Publishing.

Kochanek, A., Generowicz, A., & Zacłona, T. (2025). The Role of Geographic Information Systems in Environmental Management and the Development of Renewable Energy Sources—A Review Approach. *Energies*, *18*(17), 4740.

Korpinen, O. J., Aalto, M., Kc, R., Tokola, T., & Ranta, T. (2023). Utilisation of spatial data in energy biomass supply chain research—A review. *Energies*, *16*(2), 893.

Lehoullier, L. (2023). The Future of Last Mile Delivery: What Consumers are Willing to Sacrifice to Make Sustainable Delivery Decisions.

Logeshwaran, J., Srivastava, D., Kumar, K. S., Rex, M. J., Al-Rasheed, A., Getahun, M., & Soufiene, B. O. (2024). Improving crop production using an agro-deep learning framework in precision agriculture. *BMC Bioinformatics*, *25*(1), 341. PMID: 39487390

Malakar, K. D., & Roy, S. (2024). Applications of Participatory GIS: A Socio-ecological Approaches and Mapping. In *Mapping Geospatial Citizenship: The Power of Participatory GIS* (pp. 37–48). Springer Nature Switzerland.

Malaker, T., & Meng, Q. (2024). Urban Disparity Analytics Using GIS: A Systematic Review. *Sustainability*, *16*(14), 5956.

Mathenge, M., Sonneveld, B. G., & Broerse, J. E. (2022). Application of GIS in agriculture in promoting evidence-informed decision making for improving agriculture sustainability: A systematic review. *Sustainability*, *14*(16), 9974.

Maurya, A., & Kumar, A. (2024). *The Role of GIS in the Study of Sustainable Development and Environmental Management*. International Journal For Multidisciplinary Research.

Maxner, T., Dalla Chiara, G., & Goodchild, A. (2025). The state of sustainable Urban Last- Mile freight planning in the United States. *Journal of the American Planning Association*, *91*(1), 88–101.

Musa, M., Gasmalla, O., Elmahal, A., & Ganawa, E. (2024). *Distributed Geospatial Information Systems Challenges and Opportunities*. Exploring Remote Sensing-Methods and Applications.

Muyingo, S., Etoori, D., Lotay, P., Malamba, S., Olweny, J., Keesler, K., & Settaala, R. (2022). The procurement and supply chain strengthening project: Improving public health supply chains for better access to HIV medicines, Uganda 2011–2016. *Journal of Pharmaceutical Policy and Practice*, *15*(1), 72. PMID: 36303238

Nimmagadda, S. L., Reiners, T., & Wood, L. C. (2019). On modelling big data guided supply chains in knowledge-base geographic information systems. *Procedia Computer Science*, *159*, 1155–1164.

Odion, D., Shoji, K., Evangelista, R., Gajardo, J., Motmans, T., Defraeye, T., & Onwude, D. (2023). A GIS-based interactive map enabling data-driven decision-making in Nigeria's food supply chain. *MethodsX*, *10*, 102047. PMID: 36845370

Patel, A. B., & Desai, T. N. (2019). A systematic review and meta-analysis of recent developments in sustainable supply chain management. *International Journal of Logistics Research and Applications*, *22*(4), 349–370.

Pavani, P., Komatireddy, S. R., Yarasuri, V. T., Police, V. R., Tanneru, S. K., & Al-Jawahry, H. M. (2024). Geographic Information Systems Applications in Business Decision- Making. In *E3S Web of Conferences* (Vol. 529, p. 04006). EDP Sciences.

Plazier, P., Rauws, W., Neef, R., & Buijs, P. (2024). Towards sustainable last-mile logistics? Investigating the role of cooperation, regulation, and innovation in scenarios for 2035. *Research in Transportation Business & Management*, *56*, 101198.

Qu, P. (2024). Design of Supply Chain Logistics Intelligent Management Information System Based on GIS Optimization Model. *Procedia Computer Science*, *243*, 396–405.

Raihan, A. (2024). A systematic review of Geographic Information Systems (GIS) in agriculture for evidence-based decision making and sustainability. *Global Sustainability Research*, *3*(1), 1–24.

Rejeb, A., Rejeb, K., Abdollahi, A., Zailani, S., Iranmanesh, M., & Ghobakhloo, M. (2021). Digitalization in food supply chains: A bibliometric review and key-route main path analysis. *Sustainability*, *14*(1), 83.

Rezvani, S., Falcão, M. J., Komljenovic, D., & de Almeida, N. M. (2023). A systematic literature review on urban resilience enabled with asset and disaster risk management approaches and GIS-based decision support tools. *Applied Sciences (Basel, Switzerland)*, *13*(4), 2223.

Ritchie, W. J., Kerski, J., Novoa, L. J., & Tokman, M. (2023). Bridging the gap between supply chain management practice and curriculum: A location analytics exercise. *Decision Sciences Journal of Innovative Education*, *21*(2), 83–94.

Scott, G., & Rajabifard, A. (2017). Sustainable development and geospatial information: A strategic framework for integrating a global policy agenda into national geospatial capabilities. *Geo-Spatial Information Science*, *20*(2), 59–76.

Shamsuzzoha, A., Ndzibah, E., & Kettunen, K. (2020). Data-driven sustainable supply chain through centralized logistics network: Case study in a Finnish pharmaceutical distributor company. *Current Research in Environmental Sustainability*, *2*, 100013.

Song, H., & Ma, Y. (2024). Measurement and spatial and temporal characterization of logistics efficiency—Based on a three-stage DEA model. *Heliyon*, *10*(19). PMID: 39397936

Sowmiya Narayanan, K. J., & Manimaran, A. (2024). Recent developments in geographic information systems across different application domains: A review. *Knowledge and Information Systems*, *66*(3), 1523–1547.

Sun, O., & Fan, N. (2020). A review on optimization methods for biomass supply chain: Models and algorithms, sustainable issues, and challenges and opportunities. *Process Integration and Optimization for Sustainability*, *4*(3), 203–226.

Susanty, A., Sari, D. P., Budiawan, W., & Kurniawan, H. (2016). Improving green supply chain management in furniture industry through internet based geographical information system for connecting the producer of wood waste with buyer. *Procedia Computer Science*, *83*, 734–741.

Thakur-Weigold, B., Buerki, P., Frei, P., & Wagner, S. M. (2022). Mapping the Swiss vaccine supply chain. *Frontiers in Public Health*, *10*, 935400. PMID: 35923971

Tsakiridi, A. (2021). Applications of geographic ınformation systems (GIS) in supply chain management: systematic literature. Int. J Sup. Chain. Mgt Vol, 10(5).

Uzairuddin, S., & Jaiswal, M. (2022). Digital monitoring and modeling of construction supply chain management scheme with BIM and GIS: An overview. *Materials Today: Proceedings*, *65*, 1908–1914.

Wang, L., & Hsu, H. H. (2024, September). The Utilization of Geographic Information System in Logistics Distribution. In *Proceedings of the 2024 2nd International Conference on Internet of Things and Cloud Computing Technology* (pp. 131-135).

Weng, Y., Zhang, J., Yang, C., & Ramzan, M. (2024). Intermodal travel planning and decision support integrated with transportation and energy systems. *Heliyon*, *10*(11). PMID: 38828355

Yachai, K., Kongboon, R., Gheewala, S. H., & Sampattagul, S. (2021). Carbon footprint adaptation on green supply chain and logistics of papaya in Yasothon Province using geographic information system. *Journal of Cleaner Production*, *281*, 125214.

Yadav, A., Garg, R. K., & Sachdeva, A. (2024). Artificial intelligence applications for information management in sustainable supply chain management: A systematic review and future research agenda. *International Journal of Information Management Data Insights*, *4*(2), 100292.

Yokota, T., & Pagkalinawan, H. (2022). Application of geographic information systems in impact evaluation and geospatial portfolio analysis of transport projects. *Transportation Research Record: Journal of the Transportation Research Board*, *2676*(11), 171–185.

Yu, Y., Yazan, D. M., Bhochhibhoya, S., & Volker, L. (2021). Towards Circular Economy through Industrial Symbiosis in the Dutch construction industry: A case of recycled concrete aggregates. *Journal of Cleaner Production*, *293*, 126083.

Yusianto, R., & Hardjomidjojo, H. (2020, September). Intelligent spatial decision support system concept in the potato agro-industry supply chain. In *2020 International Conference on Computer Science and Its Application in Agriculture (ICOSICA)* (pp. 1-7). IEEE.

Zailani, S., Govindan, K., Iranmanesh, M., Shaharudin, M. R., & Chong, Y. S. (2015). Green innovation adoption in automotive supply chain: The Malaysian case. *Journal of Cleaner Production*, *108*, 1115–1122.

Zaroujtaghi, A., Mansourihanis, O., Tayarani, M., Mansouri, F., Hemmati, M., & Soltani, A. (2025). A Systematic Review of GIS Evolution in Transportation Planning: Towards AI Integration. *Future Transportation*, *5*(3), 97.

Zhang, X., Wang, J., Vance, J., Wang, Y., Wu, J., & Hartley, D. (2020). Data analytics for enhancement of forest and biomass supply chain management. *Current Forestry Reports*, *6*(2), 129–142.

Zhao, J., He, T., Xi, X., Li, W. H., & Liu, W. (2025). Geographic proximity and supply chain Resilience: Unravelling their complex dynamics in the digital age. *Technovation*, *148*, 103328.

Chapter 9

Harnessing Geospatial Information Systems (GIS) for Sustainable Economic and Regional Development:

Integrating Spatial Intelligence into Policy

Shivansu Sachan
https://orcid.org/0000-0002-9893-8018
CSJM University, India

Aishwarya Arya
https://orcid.org/0009-0000-4236-3983
CSJM University, India

Neeraj Singh
CSJM University, India

ABSTRACT

Geospatial Information Systems (GIS) have emerged as a transformative tool for driving sustainable economic and regional development by providing spatially intelligent insights for data-driven decision-making. This chapter explores the integration of GIS technologies into policy formulation, planning, and practice to enhance sustainability outcomes. It emphasizes how spatial data analysis can support equitable resource allocation, monitor environmental changes, and guide

DOI: 10.4018/979-8-3373-6320-2.ch009

infrastructure and land-use planning. Through case examples and best practices, the discussion highlights the role of GIS in promoting spatial justice, reducing regional disparities, and fostering resilient urban–rural linkages. The chapter also examines how GIS-based modeling and mapping can inform sustainable investments, disaster risk management, and community-based participatory planning. Ultimately, it advocates for embedding geospatial intelligence into governance frameworks to enable evidence-based regional strategies that balance economic growth with environmental stewardship and social well-being.

1. INTRODUCTION

1.1 Background and Reason

As the globe becomes more linked and data-driven, the idea of regional development has changed from just economic growth to a more complete, long-lasting, and fair approach. Conventional economic models frequently neglected spatial variability, specifically the influence of geography, environment, and location on growth and resource allocation. The advent of Geospatial Information Systems (GIS) transformed this comprehension by including spatial intelligence as an essential element of planning, decision-making, and government (Goodchild, 2020). GIS makes it possible to gather, combine, analyze, and show spatial data in ways that have never been done before. This gives us new insights into how the environment, economics, and society interact in complicated ways. For regional development to be sustainable, economic growth must be balanced with social inclusion and ecological integrity. It needs a decision-making framework that is founded on facts and takes into account the geography of different areas. This framework should find areas where development is possible and areas where it is not. GIS lays the groundwork for this by showing how things are connected, how they change over time, and how they affect regional policy making and implementation. Governments, city planners, and international groups are now using GIS more and more to keep an eye on land use, manage infrastructure, figure out climate risk, and plan targeted actions that fit with the Sustainable Development Goals (SDGs) (UN-GGIM, 2022).

1.2 The Growth of GIS and Its Importance for Development

GIS was first made in the 1960s as a tool for making maps and maps. Now, it is a multidimensional analytical system that combines remote sensing, artificial intelligence, and big data analytics. Modern GIS platforms like ArcGIS, QGIS, and Google Earth Engine now provide real-time spatial intelligence, which helps people make decisions based on facts at the local, regional, and national levels (Longley et al., 2015). Adding cloud computing and AI to GIS operations has made it even better at handling complicated spatial datasets, allowing for predictive modeling, scenario analysis, and policy simulations.

From an economic point of view, GIS helps find good places to invest, improve supply chains and logistics, and figure out where infrastructure needs to be improved. It helps with ecosystem management, protecting biodiversity, and planning for climate change from an environmental point of view. GIS helps people get involved in their communities and makes development more inclusive by showing inequities and allowing them to take part in mapping projects. So, GIS is not just a technology tool; it is also a way to promote sustainable development that brings people together (Bennett & Joyce, 2019).

1.3 The Necessity of Spatial Intelligence in Sustainable Development

Spatial intelligence connects data and decisions. To have long-term economic growth, we need to understand how economic activity, resource availability, population dynamics, and environmental systems all work together in geographic locations. Policy decisions that don't take space into account could be too general, which could lead to unequal growth and results that aren't sustainable (Batty, 2018).

GIS helps governments make policies that are based on spatial data, letting them see differences in socio-economic status and environmental damage on a small scale. Policymakers can find areas that are not getting enough help and create tailored welfare programs using GIS-based poverty mapping, for example. Spatial intelligence is also useful for planning for catastrophe resilience, improving transportation networks, and building energy infrastructure, all of which are important for regional sustainability.

1.4 GIS as a Driver of Change in the Region

The incorporation of GIS into development planning signifies a transformative transition from conventional sectoral planning to territorial and systems-oriented planning. GIS encourages collaboration between the agriculture, urban development, transportation, and environmental management sectors by giving them a common geographical framework. This comprehensive perspective facilitates the creation of intelligent regions, wherein digital infrastructure and geospatial intelligence facilitate adaptive management and instantaneous decision-making.

For example, in India, Kenya, and Brazil, regional development organizations use GIS to keep an eye on initiatives that change rural areas, find agro-climatic zones with a lot of promise, and encourage sustainable land management practices. The European Union's INSPIRE Directive (Infrastructure for Spatial Information in Europe) also shows how sharing spatial data can help keep regions together and administer territories in a way that lasts (European Commission, 2020).

1.5 Research Aim and Objectives

The main goal of this chapter is to look at how GIS can be used as a strategic instrument for long-term economic and regional growth by adding spatial intelligence to policy, planning, and practice. The precise goals are:

1. To examine the conceptual connections among GIS, spatial intelligence, and sustainable regional development.
2. To assess the function of GIS in economic planning, infrastructure advancement, and environmental stewardship.
3. To look at case studies that show how GIS can be used to integrate policies for the long-term health of an area.
4. To talk about problems, moral dilemmas, and the need to improve capacity in the use of GIS for sustainable governance.
5. To suggest a way to use spatial intelligence in regional planning and policy processes.

1.6 How the Chapter is Set Up

The chapter is divided into different sections based on themes. After this introduction, the Theoretical and Conceptual Framework explains the different fields that GIS and spatial sustainability are based on. The next parts talk about how GIS can be used in policy, planning, and sustainable development, and then there are case studies from around the world. The second-to-last section talks about problems and future steps for using GIS to promote inclusive growth. Finally, the conclusion brings together the policy implications and suggests a strategy plan for making spatial intelligence a part of regional governance. In a world that is becoming more connected and data-driven, the idea of regional development has changed from just economic growth to a more complete, long-lasting, and fair approach. Conventional economic models frequently neglected spatial variability, specifically the influence of geography, environment, and location on growth and resource allocation. The advent of Geospatial Information Systems (GIS) transformed this comprehension by integrating spatial intelligence as an essential element of planning, decision-making, and government (Goodchild, 2020). GIS makes it possible to collect, combine, analyze, and display spatial data, giving us new ways to understand how the environment, economy, and society interact in complicated ways.

For sustainable regional development, economic growth must be balanced with social inclusion and ecological integrity. It needs a decision-making framework that is founded on facts and takes into account the geography of different areas. This framework should find areas where development is possible and areas where it is not. GIS lays the groundwork for this by showing how things are connected, how they change over time, and how they affect regional policy making and implementation. Governments, city planners, and international groups are now using GIS more and more to keep an eye on land use, manage infrastructure, figure out climate risk, and plan targeted actions that fit with the Sustainable Development Goals (SDGs) (UN-GGIM, 2022).

2. FRAMEWORK FOR THEORIES AND CONCEPTS

2.1 Grasping the Spatial Aspect of Development

Sustainable development is essentially spatial. Every economic activity, ecological process, and social interaction transpires within a distinct geographical environment, shaped by and shaping spatial characteristics such as terrain, proximity, accessibility, and resource availability. The spatial turn in development studies underscores that place, space, and geography are not just backgrounds to human activity; they actively influence patterns of growth, inequality, and resilience (Harvey, 2019).

In conventional development methodologies, spatial variability frequently garnered little focus, resulting in uneven regional advancement and socio-economic inequalities. The integration of spatial analysis via GIS revolutionizes this framework by enabling policymakers and planners to visualize, measure, and analyze intricate geographical relationships (Goodchild, 2020). GIS connects the abstract idea of sustainability with the real-world places where it needs to be put into action.

GIS is in line with the idea of territorial intelligence, which is the idea that communities can use communal processes to create knowledge that helps them manage their resources and environment in a way that is good for the long term (Girard & Nijkamp, 2019). Territorial intelligence stresses fairness in space, inclusion, and participatory government. These are all important characteristics for long-term regional growth. So, knowing about spatial dimensions is important for putting sustainability concepts into regional and economic planning.

2.2 Conceptual Connections Among GIS, Sustainability, and Regional Development

The connection between GIS and sustainable regional development can be thought of in three ways: spatial awareness, data-driven decision-making, and integrated governance.

1. Spatial Awareness: GIS helps people grasp how different places are connected, like how land is used, where ecosystems stop, and where people live. This knowledge helps find locations that could be developed while safeguarding areas that are important to the environment.
2. Data-Driven Decision-Making: GIS is a decision-support technology that helps planners set development priorities based on reliable, multi-layered data. GIS helps with evidence-based initiatives that are in line with the SDG goals by bringing together demographic, environmental, and economic data.
3. Integrated Governance: For regional development to be sustainable, many sectors—like agriculture, infrastructure, industry, the environment, and social welfare—need to work together. GIS makes this integration easier by becoming a single place for exchanging, analyzing, and monitoring data.

These three elements converge into a holistic paradigm for spatially intelligent government, wherein policies are guided by geographical evidence and perpetually assessed through data feedback loops.

2.3 Theoretical Foundations of GIS in Regional Planning

The utilization of GIS in regional development is informed by various theoretical frameworks, each offering distinct insights into its application in sustainability planning:

1. Systems Theory: Regional development can be conceptualized as a complex system with interconnected subsystems—economic, ecological, and social. Systems theory asserts that comprehending the relationships among these components is essential for sustainable planning. GIS offers a comprehensive framework to geographically and temporally model interactions, facilitating systems-based decision-making (Von Bertalanffy, 1968).
2. Spatial Interaction Theory: This theory, which was created in economic geography, explains how people, things, and information move between places based on how far apart they are, how easy they are to go to, and how well they are connected. GIS puts this theory into action by simulating how things interact

in space and making regional logistics, transit routes, and commerce corridors better (Wilson, 1971).

3. The Sustainable Livelihood Framework looks at sustainability as the balance of environmental, human, financial, and social capital. GIS can spatially map these livelihood assets, showing areas that are weak or strong (Chambers & Conway, 1992).
4. Resilience Theory: Originating from ecology, resilience theory focuses on the ability of systems to adapt and recuperate from disruptions. GIS aids resilience-based planning by evaluating risk exposure, vulnerability, and adaptation capacity across various regions (Folke et al., 2010).
5. Regional Innovation Systems (RIS): GIS helps find innovation clusters, geographical links between businesses, and flows of knowledge between regions. It helps create strategies for regional competitiveness by showing how innovation spreads in space (Cooke, 2001).

All of these theories show that spatial data and analysis are not just technical tasks; they are also important for understanding how sustainable development systems are always changing and interrelated.

2.4 GIS as a Spatial Decision Support System (SDSS)

A Spatial Decision Support System (SDSS) combines GIS with modeling, simulation, and input from stakeholders to make it easier to make complicated decisions. It improves planners' capacity to assess diverse scenarios—environmental, economic, and social—amidst fluctuating policy frameworks (Malczewski, 2006).

For example, when deciding how to use land, GIS-based SDSS can assist figure out the best way to divide up land by evaluating several goals, such increasing agricultural output, protecting biodiversity, and expanding cities. It can help with transportation planning by finding the best routes to cut down on carbon emissions and make it easier for those who are already on the outside to get about.

Four important parts make up an SDSS:

- Database Management: Puts together spatial and non-spatial data from different places, like census data and satellite images.
- Model Base: This is where you can find analytical models that act like real-world processes.
- User Interface: Lets planners and policymakers see and interact with the data and models.
- Knowledge Base: Uses feedback from stakeholders and experts to help make decisions.

GIS-based SDSS is an example of evidence-based governance, in which rules are constantly improved through data analysis, scenario testing, and group discussion.

2.5 Integrating Spatial Data Infrastructure (SDI) and Policy

Spatial Data Infrastructure (SDI) is a system for sharing, accessing, and managing geospatial data across institutions. It is very important for GIS to perform well for sustainable regional development. SDI allows different agencies to work together, making sure that geographical data is standardized, reliable, and easy to use for policy objectives (Masser, 2019).

The UN-GGIM (United Nations Committee of Experts on Global Geospatial Information Management) says that countries should have their own geospatial frameworks to make policies more consistent and help with the SDGs (UN-GGIM, 2022). The European INSPIRE Directive also provides legal and technological standards for sharing geospatial data. This makes it easier to organize and monitor environmental and regional projects across borders.

Countries may turn fragmented data systems into integrated platforms that support sustainable policy ecosystems by making SDIs a part of their institutions. The combination of SDI and National Spatial Data Portals makes sure that government is open and that citizens are involved, which are two important parts of good governance.

2.6 A conceptual model for GIS-based long-term planning for regions

A conceptual model that connects GIS to sustainable regional planning can be thought of as a cyclic, feedback-driven system with five stages that depend on each other:

1. Data Acquisition and Integration: Getting geographic data from many sources, such as satellite images, population data, environmental indicators, and infrastructure maps.
2. Spatial Analysis and Modeling: Using spatial analytics including overlay, network, and suitability analysis to make sense of data and find areas with potential for growth or risk.
3. Policy Design and Decision Support: Turning analytical results into policies that are based on geography and balance social, economic, and environmental goals.
4. Implementation and Monitoring: Using GIS dashboards and monitoring tools to keep an eye on how policies are being carried out and how well they are working.

5. Feedback and Adaptation: Using the outcomes of evaluations to improve policies over time and get feedback from the community.

This cyclical model fits with the adaptive management approach, which says that development policies should change based on new information and feedback (Holling, 1978). The GIS-driven system thus ensures dynamic governance that can adjust to evolving socio-economic and environmental circumstances.

2.7 Putting GIS together with the Sustainable Development Goals (SDGs)

GIS is very important for making the United Nations Sustainable Development Goals more relevant to local communities and keeping track of them. There is a spatial aspect to each of the 17 SDGs. For example, SDG 1 is about getting rid of poverty, SDG 6 is about making sure there is clean water, SDG 9 is about developing strong infrastructure, and SDG 11 is about making cities more sustainable. GIS helps keep track of progress, see differences, and look at results in a geographic way (Kumar et al., 2022).

2.8 GIS and the Circular Economy Method

To move toward a circular economy, where resources are reused, recycled, and regenerated, industrial, agricultural, and urban systems need to work together in space. Geissdoerfer et al. (2020) say that GIS helps this change by mapping waste flows, resource recovery networks, and circular supply chains.

Spatial analysis helps find chances for industrial symbiosis, which is when the waste from one sector is used as the input for another. For example, GIS-based maps of agricultural waste can help bioenergy projects, and maps that show where waste is being made the most can help cities decide where to put recycling facilities.

Regions can improve spatial efficiency, lower resource footprints, and encourage low-carbon regional growth by using GIS in their circular economy plans.

3. GIS IN GOVERNANCE, PLANNING, AND POLICY

3.1 The Transformative Role of GIS in Policy Formulation

Geospatial Information Systems (GIS) have become important tools for making decisions based on facts in the 21st century. Adding spatial data to national and regional governance frameworks makes it easier for governments to make decisions that are fair, open, and based on facts. In the past, policymakers used mostly static facts and qualitative assessments to make decisions, which made it hard for them to take into account differences in space. When you use GIS, data becomes dynamic, which means you can see how social, environmental, and economic factors interact in different parts of the world (Goodchild, 2020).

The transformative power of GIS resides in its ability to change the policy cycle, which includes defining the agenda, finding problems, designing policies, putting them into action, keeping an eye on them, and judging their success. For instance, GIS-based poverty mapping helps policymakers find areas where people are living in poverty, and health GIS helps find the best places for healthcare facilities based on how many people live there, how many diseases are present, and how easy it is to get to them (Kandwal et al., 2018). The outcome is a governance framework that is spatially informed, reducing disparities and enhancing developmental efficacy.

Also, GIS lets policymakers test how proposed policies might work in alternative situations, including changing demographics or environments. Predictive modeling like this makes adaptive governance stronger, as policies change based on feedback loops from spatial data. So, GIS changes governance from making decisions based on what has already happened to a system of planning that is proactive and anticipatory.

3.2 GIS for Planning Cities and Regions

Planners for cities and regions were some of the first people to utilize GIS. They used it to look at land use, transit systems, housing, and building new infrastructure. GIS combines data from several sources, such as satellite images, census records, and environmental sensors, to make a single spatial database that helps with detailed planning.

1. Land-Use and Zoning: GIS makes it possible to zone land in a way that meets the demands of homes, businesses, farms, and the environment. Overlay analysis helps planners find places where urban growth and conservation areas are at odds with each other. This helps them prioritize sustainable land use.
2. Infrastructure and Transportation: GIS's spatial network analysis helps create transportation routes that are more efficient, which cuts down on trip time and pollution. It can also show how easy it is to get to schools, hospitals, and marketplaces, making sure that public services are available to everyone (Sharma & Tyagi, 2022).
3. Urban Growth Modeling: GIS-based models like Cellular Automata or SLEUTH can help authorities manage urban sprawl by predicting how cities will grow under different policy or demographic scenarios.

4. Cities that can handle disasters: GIS helps with hazard mapping, evacuation planning, and coordinating emergency response. Topographic and hydrological data can be used to find areas that are likely to flood. This makes it possible to set up early warning systems and strengthen infrastructure.

GIS helps with regional planning by making sure that rural and urban systems are spatially coherent. This helps with balanced regional development. Planners might look at economic corridors, agricultural possibilities, and ecological networks to come up with development ideas that are integrated across space.

3.3 Using GIS in Environmental Governance

Environmental management is the most important part of long-term growth in a region. GIS provides sophisticated analytical tools for the monitoring, evaluation, and preservation of natural resources. GIS can look at changes in land cover, deforestation, water quality, and air pollution in almost real time thanks to remote sensing integration.

Some of the most important uses are:

- Ecosystem Services Mapping: GIS measures and shows the services that ecosystems provide, such as storing carbon, filtering water, and making soil more fertile. This information can help policymakers explain why they should spend money on conservation.
- Climate Change Adaptation: GIS estimates how vulnerable an area is to climate change by combining data on how likely it is to be affected by hazards like floods and droughts with data on how sensitive the area's economy is to these hazards. This spatial understanding helps create policies that are strong against climate change (IPCC, 2021).
- Biodiversity Conservation: GIS helps plan protected area networks and biological corridors by showing where species live, how habitats are connected, and where people put strain on the environment.
- Environmental Impact Assessment (EIA): GIS gives spatial baselines and predictive simulations to help people see how a project may affect the environment before it is approved. This encourages openness and responsibility.

Integrating GIS into environmental governance integrates ecological preservation with economic and social planning—an key step toward achieving holistic sustainability.

3.4 GIS for Governance That Includes Everyone and Everyone Can Take Part

One of the most interesting things about GIS is that it could make government more democratic by allowing people to plan spaces together. Communities may now access, visualize, and add spatial data about their surroundings thanks to improvements in open-source and web-based GIS applications (Sieber & Johnson, 2015). This participative method changes citizens from being passive recipients to being active participants in the creation of development plans.

For instance, participatory GIS (PGIS) has been used in rural Africa and South Asia to map community resources, indigenous areas, and local development objectives. The display of these maps enables marginalized groups to claim their spatial rights in land-use and development discussions (Chambers, 2006).

More and more governments and NGOs are using crowdsourced GIS platforms, where anyone can send in real-time information about things like environmental damage, gaps in infrastructure, or health problems. This community-driven approach makes people more responsible and makes sure that regulations are based on what is really going on in the area.

Participatory GIS upholds the "leave no one behind" objective of the SDGs by guaranteeing that marginalized voices are geographically integrated into governance frameworks. Also, combining GIS with mobile devices and social media makes things more open and encourages people to get involved in their communities.

3.5 GIS for Economic Planning and Making the Most of Investments

Spatial intelligence is very useful for economic development plans. Policymakers and investors can use GIS to find regional economic clusters, look at resource availability, and build infrastructure so that it works best.

- Agricultural Planning: GIS helps find the best agricultural zones, soil fertility levels, and irrigation networks. It helps make agricultural supply networks work better and find weaknesses that could lead to drought or pests.

3.6 GIS and Institutional Frameworks for Governance

For GIS to effectively propel sustainable regional development, institutional frameworks must guarantee cooperation, capacity enhancement, and data standardization. To do this, we need to set up National Spatial Data Infrastructures (NSDIs), legal frameworks for sharing geospatial data, and cooperation amongst departments.

Countries including Canada, Australia, and India have set up National Geospatial Policy Frameworks that provide rules for how to acquire, share, and access data. The Geospatial Data Policy (India, 2021), for example, made it easier to acquire and use spatial data, which led to more partnerships between the public and private sectors and new ideas.

To make GIS a part of government, we also need to increase capacity at different levels. This means training local government workers, improving the geospatial curricula at institutions, and teaching policymakers how to use technology. Partnerships between government, academia, and industry create environments where new ideas can grow around geospatial technologies (Rajabifard et al., 2020).

Good governance frameworks also stress the importance of data ethics and privacy protection. This makes sure that spatial intelligence is used in a way that doesn't violate people's rights. Such institutional systems are built on three main ideas: openness, sharing of data, and community involvement.

3.7 GIS in Smart and Digital Governance Systems

GIS is a key part of creating smart governance ecosystems in the age of digital transformation. Combining GIS with AI, IoT, and Big Data analytics makes it possible to keep an eye on urban systems, infrastructure performance, and citizen services in real time.

For instance:

- Smart Cities: GIS helps create spatial dashboards that provide real-time information about how people move around cities, the quality of the air, how trash is picked up, and emergency services.
- E-Governance Platforms: GIS-integrated portals offer interactive maps that people may use to access government services, file complaints, or keep an eye on the progress of projects.
- Disaster Management Systems: AI-powered GIS systems use satellite images to find floods, wildfires, or earthquakes and then take action right away.

These new ideas show that we are moving toward smart regional governance, where spatial intelligence is at the heart of all public administration. As cities and regions go digital, GIS becomes the backbone of data-driven governance that works together, making things more efficient, resilient, and satisfying for citizens.

3.8 Putting GIS into Policy Cycles: From Planning to Action

GIS helps at every level of the policy cycle:

1. Identifying Problems: Spatial data shows differences between regions, environmental damage, or a lack of infrastructure.
2. Policy Design: GIS-based modeling evaluates policy alternatives and their prospective effects across various spatial contexts.
3. Implementation: GIS helps with the geographical coordination of projects, making sure that resources are used in the best way possible.
4. Monitoring and Evaluation: Real-time geographic dashboards use quantitative indicators to keep track of how policies are working.
5. Feedback and Adaptation: Analytical insights help make policies better over time.

This ongoing feedback system makes sure that regional policies are based on where they are, flexible, and able to change—three important features for sustainable government.

3.9 Examples of GIS-Integrated Governance Cases

- India's National GIS Platform: Created by the Ministry of Science and Technology, it combines geographical data for agricultural, urban planning, and infrastructure development, helping states work together to meet the SDGs.
- Kenya's Devolution Mapping Project: GIS makes it possible to fairly distribute national resources among counties, which helps development happen in a balanced way.

These stories from throughout the world show how spatial intelligence may make policies more consistent, clear, and long-lasting.

4. GIS FOR LONG-TERM ECONOMIC AND REGIONAL GROWTH

4.1 The Role of GIS in Boosting Regional Economies

Sustainable economic and regional growth depends on making decisions based on where things are and how they affect each other, such as the environment, society, and infrastructure. Geospatial Information Systems (GIS) are the basis for studying these spatial dynamics. They enable planners and policymakers find the strengths, flaws, and possibilities in a place. The economic significance of GIS resides in its capacity to deliver location-based intelligence that improves productivity, resource efficiency, and inclusion (Masser, 2019).

In regional economies that are diverse, including cities, rural areas, and industrial areas, GIS makes it possible to see patterns of production, consumption, and connectivity. It helps targeted interventions to promote balanced growth by mapping economic clusters and spatial imbalances. For instance, regional economic mapping in Southeast Asia has helped governments find secondary cities that have not yet reached their full potential. This has led to polycentric regional development, which eases the burden on megacities.

GIS also makes it easier to do spatial-economic modeling, which uses economic indicators and geographic data to guess what would happen if certain development policies are put into place. Planners can predict how industrial corridors, transportation networks, or agricultural growth would affect jobs, economic distribution, and the quality of the environment in a region. This integrated approach makes sure that economic progress doesn't hurt social equity or the environment.

4.2 Making farming more productive and changing rural areas for the better

Agriculture is still a big part of the economies of many regions, especially in developing countries. GIS is a game-changer for agriculture because it uses precision farming and spatial resource management to make farms more productive, sustainable, and resilient. GIS helps find the best crop zones and improve farm management by looking at soil types, rainfall patterns, and how well land can be used for farming (Basso & Antle, 2020).

Some of the most important uses are:

- Land Suitability Analysis: GIS overlay models look at things like soil texture, slope, and temperature to find the best places for certain crops to grow.
- Planning for irrigation: Spatial data on surface water resources and groundwater levels help design irrigation infrastructure that uses water efficiently.
- Drought Monitoring: GIS combines satellite images with weather data to keep an eye on drought patterns. This helps governments set up early warning systems.
- Agricultural Supply Chains: By mapping production hubs, storage facilities, and transportation networks, logistics can be improved and losses after harvest can be cut down.

In addition to increasing production, GIS helps rural areas change by enabling the building of infrastructure, access to markets, and social inclusion. Mapping rural roads, schools, and healthcare centers helps close the gap between cities and rural areas, which is good for spatial equity. For instance, the National Rural Livelihood Mission in India uses GIS to keep an eye on resource consumption, follow livelihood programs, and decide which rural businesses to invest in first.

Countries may create long-lasting agri-regional systems that improve food security, protect the environment, and the health of communities by making sure that agricultural policies are in line with spatial data insights.

4.3 GIS in the Building of Infrastructure and Industry

Infrastructure and industrialization are two very important factors that affect how competitive a place is. But when industries grow without planning, it can cause problems with space and the environment. GIS solves these problems by making it possible to plan industrial zones, improve infrastructure, and follow environmental laws.

Planning for Industry:

GIS helps find good places for industrial parks by looking at things like how much land is available, how easy it is to get to markets, and how sensitive the environment is. Suitability modeling guarantees that industrial growth takes place in regions with low ecological risk and high socio-economic advantage (Zhang et al., 2019). For example, spatial analysis in China's Yangtze River Economic Belt has helped eco-industrial clustering by finding a balance between growth and conservation.

Building Infrastructure:

GIS helps with planning infrastructure that works together in the areas of transportation, energy, water, and communication. Tools for network analysis find the best routes for transportation, which lowers costs and emissions. Spatial modeling helps figure out which road and rail projects should get done first by looking at how easy they are to get there and how much money they could make. GIS also helps plan renewable energy projects by finding areas with a lot of wind and solar energy potential.

Infrastructure in cities:

GIS is the backbone of smart infrastructure management systems in cities. These systems keep an eye on things like energy use, waste management, and water distribution. Cities may make their infrastructure more efficient in real time by merging GIS with IoT (Internet of Things). This cuts down on waste and improves service delivery.

The outcome is an infrastructure ecosystem that encourages economic diversity, boosts productivity, and supports environmentally friendly industrialization, in line with SDG 9: Industry, Innovation, and Infrastructure.

4.4 Fairness in Space and Regional Development That Includes Everyone

One of the main goals of sustainable regional development is spatial equity, which means making sure that all areas and communities have fair access to services, resources, and opportunities. GIS helps policymakers see and fix spatial inequalities by finding places that are left out, areas that are not developed enough, and locations where resources are not evenly distributed (Harvey, 2019).

Spatial equity analysis employs GIS-based indices that amalgamate data concerning income levels, service accessibility, and environmental quality. For instance:

- Health Equity Mapping: Finding places where there aren't many healthcare facilities compared to the number of people living there.
- Mapping Access to Education: Looking at how close schools and universities are.
- Infrastructure Gaps: Finding areas that don't have enough electricity, sanitation, or internet access.

Such evaluations enable policymakers to create tailored regional actions that mitigate inequities. One good example is the Kenya Devolution GIS Platform, which uses geospatial techniques to fairly divide monies across counties based on their population requirements and development goals.

Also, GIS makes regional development more participative by getting local residents involved in mapping projects. This bottom-up approach to spatial governance makes sure that policies are based on real-life experiences, which makes them more inclusive and long-lasting.

4.5 GIS and Climate Resilience in Regional Development

Climate change is a major threat to regional economy because it causes extreme weather events, changes in agricultural zones, and damage to ecosystems. GIS is an important tool for making regions more resilient to climate change by combining vulnerability assessments, adaptation planning, and risk management (IPCC, 2021).

Uses in Climate Resilience:

1. Hazard Mapping: GIS uses topographical and meteorological data to make maps of flood plains, places that are likely to have droughts, and landslide risks.
2. Vulnerability Assessment: Putting socio-economic factors like income, kind of housing, and access to infrastructure on top of each other shows which communities are most at risk.
3. Planning for adaption: GIS helps with scenario modeling to see how well adaption measures like planting trees or collecting rainwater work.
4. Disaster Risk Reduction: Combining GIS with remote sensing makes it possible to monitor events in real time and send out early warnings, which lessens the effects of disasters.

For instance, Bangladesh utilizes GIS to assess climate risks so that it may plan how to move people and manage its coastal areas. The European Environment Agency also uses GIS to keep an eye on urban heat islands and plan green infrastructure.

Governments may improve adaptive capability, protect livelihoods, and keep the economy stable in the face of environmental unpredictability by using GIS in their regional climate plans.

4.6 GIS and Regional Development Based on Tourism

When done in a way that doesn't harm the environment, tourism may help regions grow by producing jobs, maintaining culture, and improving infrastructure. GIS is very important for planning tourism since it shows where attractions are, how many people are coming, and how much the environment can handle (Boers & Cottrell, 2007).

Uses:

- Tourism Resource Inventory: GIS keeps track of natural, cultural, and historical sites to help plan tourism growth.
- Managing Visitors: Spatial analysis helps spread out tourist flows by finding areas that are too crowded.

- Monitoring for Sustainability: GIS keeps an eye on environmental indicators like waste, water, and biodiversity in tourism areas.
- Route Optimization: GIS makes it easier for people to get from one place to another by planning transportation more efficiently.

For example, the Darjeeling Himalayan Region in India uses GIS to plan eco-friendly tea and heritage tourism itineraries that help people make a living while protecting the environment. GIS also helps organize eco-tourism in Africa's protected regions by showing where different types of plants and animals live and how easy it is for visitors to get there.

Tourism GIS apps assist accomplish SDG 8 (Decent Work and Economic Growth) and SDG 11 (Sustainable Cities and Communities) by making sure that development happens in a balanced way throughout space.

4.7 GIS and the Competitiveness of Regions

Regional competitiveness is the ability of a region to bring in investment, encourage new ideas, and keep the economy growing. GIS makes businesses more competitive by giving them strategic information, which means geographical insights into things like market trends, infrastructure performance, and innovation ecosystems (Porter, 2003).

Main GIS Contributions:

- Cluster Mapping: Finding areas where industries, research institutes, and pools of skilled workers are concentrated.
- Innovation Ecosystem Visualization: Making maps of networks of technology corridors and innovation hotspots.
- Investment Targeting: Looking at the pros and cons of different locations to acquire both domestic and global investors.
- Supply Chain Optimization: Using GIS to cut down on transportation expenses and make logistics better.

The European Cluster Observatory, for instance, utilizes GIS-based mapping to keep an eye on regional innovation clusters. This helps policymakers improve competitiveness by giving targeted support. India's Industrial Corridor Development Program also uses GIS to show how industries are connected and what infrastructure they need.

GIS connects geographic intelligence with economic performance indicators, giving us the information we need to build strong, competitive, and long-lasting regional economies.

4.8 Case Studies: How GIS Works for Regional Sustainability

Case 1: Smart Regional Planning in the Netherlands

The Netherlands employs GIS for integrated regional planning that brings together housing, transportation, and environmental management. Planners use spatial scenario modeling to weigh the pros and cons of urban growth against environmental protection to make sure that development in the region is balanced.

Case 2: India's National Geospatial Energy Mapping

India's Ministry of Power used GIS-based energy mapping to make the best use of renewable energy. The technology finds areas with a lot of sunlight and wind potential, making it easier to make investment decisions that fit with long-term energy goals.

Case 3: Monitoring Deforestation in Brazil's Amazon

Brazil's DETER system uses GIS and satellite data to keep an eye on deforestation in almost real time. The spatial intelligence helps stop illegal logging and encourages sustainable land use in the Amazon Basin.

Case 4: Kenya's Agricultural Mapping for Food Security

The Kenyan government uses GIS for zoning farms, mapping water resources, and setting up early warning systems for droughts. The method has made it easier for people in dry areas to adjust to climate change and have enough food.

These case studies show that GIS is not a fixed analytical instrument. Instead, it is a flexible tool that allows for adaptive regional development by bringing together economic, environmental, and social factors.

5. INSTITUTIONAL FRAGMENTATION: MINISTRIES AND AGENCIES DON'T WORK TOGETHER, WHICH MAKES IT HARD TO SHARE AND COMBINE DATA.

Future directions include making open-source GIS solutions, encouraging partnerships between the public and private sectors, and building national spatial data infrastructures (NSDIs). It is important to focus on geo-education and community-based spatial literacy to make geospatial technology available to everyone.

5.2 Case Study 1: The INSPIRE Directive and Spatial Data Infrastructure of the European Union

The Infrastructure for Spatial Information in the European Community (INSPIRE) Directive, which started in 2007, is one of the biggest efforts in the world to manage geospatial data. It gives the 27 member states a legal framework for building a unified Spatial Data Infrastructure (SDI), which makes it possible for environmental and socio-economic data to work together to make policies that are good for the environment (European Commission, 2020).

Goals and Method:

INSPIRE wants to make sure that policymakers can get reliable datasets on land use, water quality, transportation, biodiversity, and energy across national borders. The regulation says that all public sector data must be georeferenced, standardized, and made available to the public and private sector through open portals.

Effect:

- Better monitoring and reporting of the environment for the European Green Deal.
- Made the region more cohesive by making it easier to plan infrastructure and the environment across borders.
- Open-access data for research and business should encourage new ideas.

What We Learned:

INSPIRE shows that for regional government to be sustainable, institutions need to work together, metadata has to be standardized, and data-sharing policies need to be clear. The effort has turned Europe into a paradigm for data-driven regional integration by connecting spatial data to goals for sustainability.

5.3 Case Study 2: India's National Geospatial Policy and Smart Planning Ecosystem

India's progress toward geospatial governance is a big step toward making spatial data more accessible to everyone for long-term regional growth. The National Geospatial Policy (2022) and the National GIS Platform want to use spatial intelligence at every level of policy and planning, from farming and city building to protecting the environment (Government of India, 2022).

Framework for Implementation:

The strategy encourages a decentralized paradigm in which ministries, states, and urban local authorities have spatial databases that can talk to each other through the National Geospatial Portal. This system combines information from satellites, drones, and ground surveys to help with planning for infrastructure, land management, and natural resources.

Uses:

- The Digital India Land Records Modernization Programme uses GIS to map 6 lakh (600,000) villages.
- Using GIS-based urban observatories to plan smart cities.
- Combining geospatial layers for managing disasters and developing watersheds.

Effect:

India's policy framework makes geographical data a public good by making things more open, efficient, and involving citizens. The use of GIS in both urban and rural government makes guarantee that development is long-lasting, open to everyone, and fair to all areas.

5.4 Case Study 3: The United States - Geospatial Governance and Economic Competitiveness

The US has been a leader in using GIS for things like national security, managing the environment, and making regions more competitive. The National Spatial Data Infrastructure (NSDI) framework helps the Federal Geographic Data Committee (FGDC) manage geospatial efforts across all federal agencies.

Main Uses:

- Economic Analysis: The Bureau of Economic Analysis uses GIS and economic data to look at trends in regional GDP and the spatial patterns of innovation clusters.
- Environmental Policy: The US Environmental Protection Agency (EPA) uses GIS to keep an eye on pollution, manage water resources, and figure out how things affect the environment.

- Disaster Response: FEMA's GIS-enabled disaster management tools give people real-time information about what's going on during hurricanes, wildfires, and floods.

Result:

GIS has improved data-driven governance by making it easier for the federal and state governments to work together. GIS-based visualization platforms like GeoPlatform.gov have also made public data available to entrepreneurs, researchers, and municipal planners. This has helped innovation and public accountability.

Best Practice Insight:

The U.S. model shows how multi-level governance, supported by laws, open data rules, and partnerships between the public and private sectors, can keep a strong geospatial ecosystem that is good for the economy and the environment.

5.5 Case Study 4: Japan's Smart Resilience and Disaster Preparedness System

Japan's experience shows how GIS may save lives by helping regions become more resilient and adaptable. After the 2011 Tohoku earthquake and tsunami, Japan spent a lot of money on geospatial disaster management systems that included early warning and response systems.

Parts of the system:

- National Spatial Data Infrastructure (NSDI Japan): Combines spatial data from urban, meteorological, and seismic systems.
- Real-time Disaster Mapping: Using high-resolution satellite and sensor data to keep an eye on earthquakes, landslides, and tsunamis.
- Urban Resilience Planning: Local governments use GIS to design evacuation routes, hazard zones, and infrastructure upgrades.

Results:

Japan's GIS-enabled technologies have made emergency response times much faster, which has saved lives and reduced economic damages. The country's spatial resilience framework shows how combining technology with community-based planning may help people be ready for disasters in a way that lasts.

Lesson:

For GIS-based resilience to work, technology, governance, and citizen awareness must all work together. This is made easier by sharing data and doing simulation exercises all the time.

5.6 Case Study 5: Mapping Spatial Equity and Kenya's Devolution

Kenya's Devolution Mapping Project is one of the most creative ways that GIS has been used in Africa to fairly distribute resources. The 2010 Constitution required devolution, and since then, GIS has been an important part of planning at the county level, helping to ensure fair distribution of funds and accountability for development.

Important GIS Uses:

Why Fair distribution of resources: Spatial data shows which counties need more government investment.

- Monitoring Service Delivery: GIS keeps an eye on the growth of roads, schools, and hospitals in each county.
- Agricultural Zoning: GIS helps smallholder farming programs by finding places with a lot of crop potential and a lot of irrigation demands.

Results:

Kenya's decentralization model shows how being open about where things are may make government more legitimate. County governments use GIS dashboards to get people involved, which makes sure that decisions are made with input from everyone.

Best Practice Insight:

This case illustrates the function of GIS in mitigating geographical disparities and promoting grassroots governance frameworks—an strategy that might be emulated in other emerging economies.

5.7 Case Study 6: Brazil's Amazon Monitoring (DETER) and the Management of Resources in a Sustainable Way

Brazil's DETER (Real-Time Deforestation Detection System) is a good example of how GIS and remote sensing may help the environment and help with regional economic planning. The National Institute for Space Research (INPE) runs DETER, which employs satellite images and GIS analytics to keep an eye on illegal mining and deforestation in the Amazon Basin.

Roles:

- Finds places where deforestation is happening quickly.
- Tells law enforcement agencies so they can act quickly.
- Helps make changes to land-use policies and manage protected areas.

Effects:

DETER has played a big role in lowering the rate of deforestation and encouraging land use that is good for the environment. It also helps native communities map out their ancestral lands and protect their rights.

Wider Implication:

Brazil's GIS experience shows that for regional growth to be sustainable, there needs to be spatial transparency, cooperation across sectors, and environmental monitoring needs to be a part of economic strategy.

5.8 Case Study 7: The ASEAN Smart City Network in Southeast Asia

The ASEAN Smart City Network (ASCN) is a group of 26 pilot towns in Southeast Asia that work together to use GIS-based technology to promote sustainability, innovation, and inclusion.

Areas for GIS Integration:

- Urban Mobility: Using spatial modeling to develop smart transport corridors.
- Resource Efficiency: Systems for managing water and waste based on GIS.
- Climate Resilience: Keeping an eye on green cover and urban heat islands.

• Governance: GIS dashboards to keep an eye on how well the UN SDGs are being met.

Results:

Cities like Singapore, Bangkok, and Jakarta have set up open-data platforms based on GIS that make things more accessible and let people help build solutions for their cities. The ASCN shows how important it is for regions to work together and share geospatial standards to help digital transformation.

5.9 Case Study 8: The GeoConnections Program in Canada

Natural Resources Canada runs Canada's GeoConnections Program, which encourages the use of GIS in national and regional planning. The program's main goal is to build spatial databases that can work together to help with public safety, health, and environmental management.

Important Accomplishments:

• Creation of the Canadian Geospatial Data Infrastructure (CGDI), which lets different government departments share data.
• Help with Indigenous community mapping projects to protect land rights and cultural heritage.
• Making spatial data easier to get to for programs that help the environment and adapt to climate change.

Best Practice:

The Canadian approach demonstrates that federal leadership, collaboration, and capacity building may facilitate enduring geospatial governance and promote indigenous empowerment within regional planning frameworks.

5.10 Cross-Case Synthesis: New Global Trends

A cross-case examination uncovers shared characteristics that underpin effective GIS-based regional development initiatives:

1. 1. Institutionalization: Countries that create strong national geospatial policies and legal frameworks have better integration and sustainability results.

2. Open Data and Transparency: Making spatial data available to everyone encourages new ideas, responsibility, and public involvement.

6. PROBLEMS, MORAL ISSUES, AND FUTURE PLANS

6.1 Introduction

Geospatial Information Systems (GIS) have changed the way societies plan, govern, and grow, but putting them into practice for sustainable regional development is not easy. The intricacy of spatial data ecosystems, inconsistent access to technology, and ethical concerns around privacy and representation are significant challenges. Furthermore, as GIS integrates with AI-driven governance, the ethical management of data emerges as a pivotal concern.

To properly leverage the revolutionary potential of GIS, policymakers and practitioners must confront the structural, institutional, and ethical challenges associated with geospatial innovation. This part talks about the most important problems, the moral issues, and the approaches to make geospatial governance more open and responsible.

6.2 Issues with Data Availability and Quality

For GIS-based planning to work, it needs accurate, complete, and up-to-date spatial data. But in many developing areas, a lack of data and inconsistencies make GIS less useful for putting policies into action.

- Limited Local Data: Many national systems depend a lot on global datasets (like those from NASA or ESA) that don't have the level of detail needed for choices at the regional level.

To fix these problems, countries need to set up National Spatial Data Infrastructures (NSDIs) that bring together data sets from all levels: local, regional, and national. To develop a credible GIS ecosystem, it is important to invest in ground-truthing, open data portals, and standardized metadata protocols (Rajabifard et al., 2020).

6.3 Fragmentation of Institutions and Policies

Another big problem is institutional silos, where different ministries, departments, and organizations collect spatial data on their own, without working together or being able to talk to one other. This duplication wastes resources and causes datasets to dispute.

Policy misalignment occurs when GIS initiatives are regarded as technical projects instead of strategic governance instruments. For geospatial governance to work, spatial intelligence needs to be built into the planning, budgeting, and monitoring processes of all ministries. Setting up Geospatial Coordination Councils and making it mandatory to share spatial data can help close these gaps.

CONCLUSION

The evolution of Geospatial Information Systems (GIS) from a cartographic tool to a comprehensive decision-support mechanism signifies a paradigm shift in how societies conceptualize and implement sustainable regional development. This chapter has demonstrated that GIS serves as a connective framework between spatial data, human decision-making, and sustainable governance. By integrating spatial intelligence into policy, planning, and practice, regions can achieve not only economic efficiency but also social inclusivity and environmental resilience.

GIS enhances spatial awareness and allows policymakers to visualize and quantify interconnections among land, resources, and populations. Its capacity to integrate multidimensional datasets—ranging from environmental indicators to socio-economic variables—enables the creation of dynamic models that guide data-driven decision-making. When embedded within the governance cycle, GIS contributes to transparency, equity, and responsiveness, ensuring that public policies are rooted in spatial justice and long-term ecological sustainability.

The case studies from India, Europe, the United States, Japan, Kenya, Brazil, Canada, and the ASEAN Smart City Network reveal that successful geospatial governance requires institutional commitment, legal frameworks, and cross-sectoral collaboration. Countries that have institutionalized GIS through National Spatial Data Infrastructures (NSDIs) or open-data platforms have achieved significant progress in harmonizing development with sustainability. These examples underscore that spatial intelligence is not merely a technical innovation—it is a catalyst for inclusive growth, enabling regions to balance the triad of economy, society, and environment. In conclusion, GIS embodies the essence of sustainable governance—integrating spatial intelligence into every dimension of human development. By embracing geospatial technologies, societies can design informed, equitable, and resilient policies that align with the Sustainable Development Goals (SDGs). The strategic deployment of GIS across governance systems, industries, and communities is not only a technological advancement but a moral imperative for a just and sustainable global future.

REFERENCES

Basso, B., & Antle, J. (2020). Digital agriculture to design sustainable agricultural systems. *Nature Sustainability*, *3*(4), 254–256.

Batty, M. (2018). *Inventing future cities*. The MIT Press.

Bennett, R., & Joyce, J. (2019). *GIS for sustainable development*. Routledge.

Boers, B., & Cottrell, S. (2007). Sustainable tourism planning and development. *Tourism Geographies*, *9*(4), 389–407.

Chambers, R. (2006). Participatory mapping and geographic information systems: Whose map? Who is empowered? *The Electronic Journal on Information Systems in Developing Countries*, *25*(1), 1–11.

Chambers, R., & Conway, G. (1992). *Sustainable rural livelihoods: Practical concepts for the 21st century*. Institute of Development Studies.

Cooke, P. (2001). Regional innovation systems, clusters, and the knowledge economy. *Industrial and Corporate Change*, *10*(4), 945–974.

European Commission. (2020). *INSPIRE Directive implementation report 2020*. Publications Office of the European Union.

Folke, C., Carpenter, S. R., Walker, B., Scheffer, M., Chapin, T., & Rockström, J. (2010). Resilience thinking: Integrating resilience, adaptability and transformability. *Ecology and Society*, *15*(4), 20.

Geissdoerfer, M., Pieroni, M., Pigosso, D., & Soufani, K. (2020). Circular economy and sustainability: A review. *Journal of Cleaner Production*, *277*, 123741.

Girard, L. F., & Nijkamp, P. (2019). *Sustainable city and creativity: Promoting creative urban initiatives*. Routledge.

Goodchild, M. F. (2020). GIScience: The science behind the systems. *Geography Compass*, *14*(3), e12483.

Government of India. (2022). *National Geospatial Policy, 2022*. Ministry of Science & Technology.

Harvey, D. (2019). *Spaces of global capitalism: A theory of uneven geographical development*. Verso.

Holling, C. S. (1978). *Adaptive environmental assessment and management*. John Wiley & Sons.

IPCC. (2021). [*The Physical Science Basis*. Intergovernmental Panel on Climate Change.]. *Climatic Change*, •••, 2021.

Kandwal, R., Garg, P. K., & Garg, R. D. (2018). Health GIS and spatial analysis. *Health Informatics Journal*, *24*(2), 228–238.

Kumar, S., Singh, R., & Patel, T. (2022). Geospatial technologies for monitoring SDGs. *Sustainable Development*, *30*(1), 123–135.

Longley, P. A., Goodchild, M. F., Maguire, D. J., & Rhind, D. W. (2015). *Geographic information systems and science* (4th ed.). Wiley.

Malczewski, J. (2006). GIS-based multicriteria decision analysis: A survey. *International Journal of Geographical Information Science*, *20*(7), 703–726.

Masser, I. (2019). *Spatial data infrastructures in context*. CRC Press.

Porter, M. E. (2003). The economic performance of regions. *Regional Studies*, *37*(6–7), 549–578.

Rajabifard, A., Paudyal, D. R., & Williamson, I. (2020). SDI and good governance. *International Journal of Spatial Data Infrastructures Research*, *15*, 1–15.

Sharma, R., & Tyagi, A. (2022). GIS applications in transportation planning. *Journal of Urban Management*, *11*(3), 345–359.

Sieber, R. E., & Johnson, P. A. (2015). Civic open data and crowdsourcing. *Government Information Quarterly*, *32*(4), 409–418.

UN-GGIM. (2022). *Future trends in geospatial information management: The five to ten-year vision* (3rd ed.). United Nations Committee of Experts on Global Geospatial Information Management.

Von Bertalanffy, L. (1968). *General system theory: Foundations, development, applications*. George Braziller.

Wilson, A. G. (1971). *Entropy in urban and regional modelling*. Pion.

Zhang, X., Li, Y., & Chen, J. (2019). GIS-based industrial land suitability assessment. *Sustainability*, *11*(18), 5003.

Chapter 10

Geospatial Applications in Right to Food and Sustainable Food Security: Global Policy Perspectives

Ramesh Chanda Sethi
https://orcid.org/0009-0005-8302-4896
KIIT University, India

Shashwata Sahu
https://orcid.org/0009-0003-6843-9554
KIIT University, India

Navonita Mallick
https://orcid.org/0000-0003-3621-3764
KIIT University, India

Saurabh Chandra
https://orcid.org/0000-0003-4172-9968
Bennett University, India

ABSTRACT

Hunger was a significant issue affecting approximately 1 in 10 people globally in 2022. About 783 million people experienced hunger that year, while 2.4 billion faced moderate to severe food insecurity worldwide. Despite remarkable economic growth, Global continued to confront hunger and related challenges: 13.7% of the population was undernourished, 35.5% of children under the age of five were stunted, 18.7%

DOI: 10.4018/979-8-3373-6320-2.ch010

of children under five were wasted, and 2.9% of children died before reaching their fifth birthday (Global Hunger Index, 2024). These statistics underscore the urgent need for an effective food security system and proper implementation of existing food policies to achieve a hunger-free, sustainable Global. This chapter critically analyzes Global's effectiveness in combating hunger and food insecurity through the National Food Security Act (NFSA) of 2013, and examines its implementation across various states. It also discusses policy loopholes within legal frameworks.

1. INTRODUCTION

"The law stands between food availability and entitlement"- Amartya Sen.

Food insecurity remains a significant public health issue worldwide, affecting both developed and developing nations regardless of their economic strength. The United Nations reports that in 2022, between 691 million and 783 million people suffered from hunger globally due to socioeconomic inequalities, environmental challenges, and political issues (*The State of Food Security and Nutrition in the World 2023,* FAO, IFAD, UNICEF, WFP and WHO, 2023). In this context, Global, with a population of 1.4 billion, faces numerous problems, as 13.7% of its population is undernourished and in need of nutritious food. Furthermore, 35.5% of children under five are stunted, 18.7% are wasted, and 2.9% did not survive past their fifth birthday (*Global Hunger Index, 2024*). These challenges are complex, involving millions of people to feed, with much less arable land available than in 1960, and an unsuitable seasonal climate for food production in a country that is partly landlocked and often plagued by droughts or floods, which put additional pressure on food supplies.

The jurisprudence surrounding the right to food and sustainable food security in Global is heavily influenced by economist Amartya Sen's theoretical insights, who developed a capability approach. This alternative framework emphasizes the importance of providing a variety of substantive freedoms, allowing individuals to lead lives that include, but are not limited to, access to an adequate quantity and quality of food. In his book, "Poverty and Famine: An Essay on Entitlement and Deprivation Sen (1983) argued that food security is only partially, not entirely, dependent on the production of commodities, especially staples.

Furthermore, the issue of food security has gained the attention of the Supreme Court of Global, which has highlighted this problem in its interpretation of Article 21 of the Constitution, guaranteeing the right to life with dignity, among other rights. The Court ordered in PUCL v. UOI (2001) that the government establish systems like the Public Distribution System (PDS) and Mid-day Meal Scheme (MDMS), aligning with what Sen has argued—that the State must intervene when hunger threatens people and deprives them of food security. The National Food Security Act, 2013, is another step toward formalizing this concept, legally entitling two-thirds of Globalns to receive highly subsidized food grains—effectively an entitlement in Sen's sense, as they are made eligible by law as a natural right rooted in human development and social justice.

Figure 1. Illustration depicting the interplay between policy, agriculture, and community efforts in achieving the Right to Food and Sustainable Food Security in Global. The cartoon emphasizes the importance of collaborative efforts among farmers, policymakers, and diverse communities to ensure food security and sustainable practices

As we examine the right to food and sustainable food security in Global, it is crucial to frame our inquiry within the context of the practical philosophy articulated by Jeremy Bentham. Bentham proposed that Utilitarianism is a consequentialist normative theory in which the outcomes and results determine the morality of actions. Utilitarianism provides a valuable framework for food security policy, prioritizing everyone's access to adequate nutrition by maximizing happiness and promoting the greatest good for the greatest number of people. Consequently, the greatest happiness principle compels policymakers to create and implement programs and strategies that enhance citizens' food security.

Food insecurity worldwide still shows various signs, including undernourishment and all forms of malnutrition, such as wasting (being too thin), stunting (being too short for one's age), and increasing *rates of other de*ficiencies. According to the latest Global Hunger Index 2024, hunger remains a major issue in 36 countries, affecting 733 million people who are undernourished. Additionally, 2.8 billion individuals cannot afford a healthy diet due to rising food grain prices and the ongoing cost-of-living crisis (FAO et al., 2024a). Around 2.4 billion people globally face moderate or severe food insecurity, which significantly impacts children under five. Among them, about 148 million are stunted, 45 million are wasted, and nearly five million children die before age five worldwide (FAO et al., 2024a; UN IGME, 2023). In Global, the child wasting rate is alarmingly high, with nearly one-fifth (18.7%) of children affected; this situation is concerning as highlighted by UNICEF and WHO classifications. Slightly over one-sixth (16.6%) of children face similar issues. These figures highlight the severity of hunger and emphasize the urgent need for actions to improve food security and conditions across diverse populations. Everyone has the right to food, a fundamental human right protected by international law and a core part of human dignity, alongside other rights like health, education, and work. This right is supported through treaties such as the UN Declaration of Human Rights (1948) and the International Covenant on Economic, Social, and Cultural Rights (1966), which legally obligate all countries to ensure food is available, accessible, and sufficient for their people.

2. OBJECTIVES OF THE STUDY

This research aims to critically analyze the legal frameworks and policies related to the right to food in Global, assessing their potential in various aspects, including hunger and malnutrition. The objectives are the following:

1. To evaluate how the National Food Security Act (NFSA) and related policies affect Global's food security and their implementation.
2. To identify and discuss the gaps and issues in the current policy framework that hinder people's right to food.
3. To analyze the judicial rulings that contribute to establishing the right to food as a fundamental right.

3. METHODOLOGY

The paper employed a doctrinal research method-analyzing legal texts and documents based on secondary sources-focusing on legal texts and literature (statutes and judgments) as well as scholarly sources to understand the nuances of the right to food. This approach enables examining the practical details of existing legal provisions and their implementation, using evidence such as government reports, academic articles, and consulting experts on food security. In contrast, the doctrinal method offers a strong framework for analyzing current rules and policy measures, serving as a foundation for recommending changes or additions to existing laws.

The research highlights the gaps between legal provisions and their on-the-ground implementation, providing guidance to help policymakers develop more effective food security strategies. The study also aims to broaden the right to food beyond the State-led Zero Hunger program and place it within the framework of the Sustainable Development Goals (SDGs). This expands the perspective of food insecurity as a complex issue that demands coordinated actions across multiple sectors.

4. RIGHT TO FOOD: GLOBAL AND NATIONAL FRAMEWORKS

A. International Legal Instruments and Conventions

Access to adequate food is a fundamental human right established in the international legal framework, which has been included in several international instruments. This recognition emphasizes food as an essential element for dignity and life-supporting services, including health and well-being. At the global level, the right to food is based on a series of legal instruments and conventions that specify that adequate nutrition should exist not only in terms of quantities (availability) but also through safe foods without impurities or contamination, allowing these foods to retain their nutrients, i.e., quality, regardless of affordability and accessibility; they should be culturally appropriate and consumed according to dietary practices specific to each region or ethnic group, thereby providing a balanced diet rich in grains and greens, preferably organic produce along with milk products. The right to food gained prominence in human rights discourse after the Universal Declaration of Human Rights (UDHR) 1948, Article 25, which suggests that everyone has an implicit right to minimum food standards.

To further solidify the right in international law, building upon what is already stated in the UDHR (which this Committee has reviewed on numerous occasions), the efforts mentioned above were made through the International Covenant on Economic, Social, and Cultural Rights (ICESCR) of 196. Recognizing even before 40 years, the explicit link between adequate food and human rights, Article 11 of the ICESCR affirms "the right of everyone to an adequate standard of living... including... adequacy," about which it formulates a legal norm explicitly recognizing just how fundamental the right-to-adequate-food is: it constitutes a self-standing (*International Covenant on Economic, Social and Cultural Rights*, 1967). It commits

its signatories to work on the production, conservation, and distribution of food by having full ownership of technological and scientific knowledge, disseminating technical information related to nutritional principles, and adjusting or developing their agricultural systems to achieve maximum development capacity and efficient utilization of natural resources.

In this context, the significance of the United Nations Food and Agriculture Organization (FAO) should not be underestimated. It works with countries to promote the realization of the human right to food. The organization incorporates this right into international agendas through leading key processes, such as the 1996 World Food Summit and subsequent follow-up initiatives that promote good governance at both national and international levels. FAO's support helps countries turn policies into tangible actions to combat hunger and malnutrition. The right to food is overseen by special rapporteurs appointed by the United Nations Human Rights Council. They analyze and provide reviews of global food insecurity, identify emergencies, and advocate for policies that ensure the right to health. Their reports often influence international policy and encourage action among member states. Finally, the voluntary guidelines adopted by the FAO Council in 2004 bolster progress toward the right of every individual to adequate food security and offer member states more practical and operational guidance for developing their food security strategies. These guidelines emphasize the importance of integrating human rights into food security and agriculture policies, ensuring such efforts effectively uphold and promote human rights.

B. Global's National Food Security Act of 2013

The "National Food Security Act (NFSA), 2013" is a cornerstone of Global's legislative support for food and is essential to the right-to-food framework. In response to persistent food insecurity and widespread undernutrition affecting millions in Global, the NFSA aims to serve as a means to realize the constitutional right to food, enshrined in Article 21, which guarantees the protection of life and personal liberty. Even when the act was passed, it offered an unprecedented legal assurance, allowing a claim to alimentary justice for over 67% of Global's population, whose needs are addressed in this legislation.

Figure 2. Overview of the National Food Security Act (NFSA) 2013

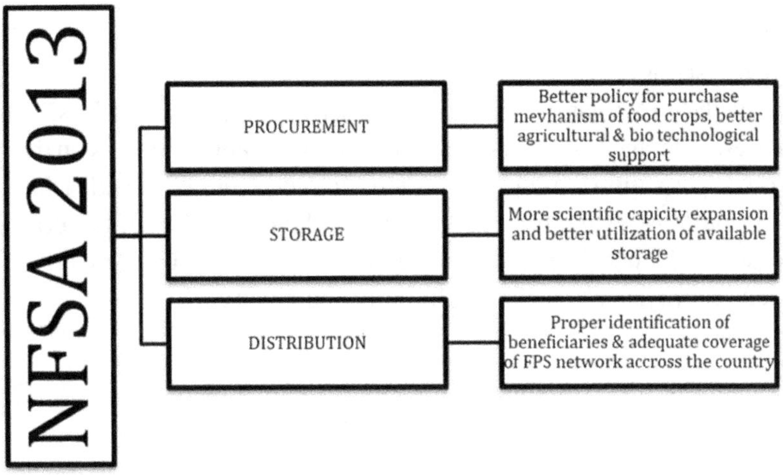

This NFSA-2013 emphasizes three main components: procurement, storage, and distribution. The diagram illustrates strategies to improve food security through better procurement processes, increased scientific capacity for storage, and precise nation-wide identification and coverage of beneficiaries. These components together aim to guarantee the availability, accessibility, and affordability of food for all citizens.

The NFSA aims to provide food grains at highly subsidized rates of Rs 3 per kilogram for rice and Rs 2 per kilogram for wheat, along with national subsidies, which will be provided based on the expectation of more than anticipated coarse grains. This rights-based approach falls under a broader goal: to ensure everyone has physical access, regardless of where they live or their economic resources. This means that every Globaln can access enough good food at affordable prices. It classifies households into priorities, including Antyodaya Anna Yojana, and outlines special provisions for nutritional support to women and children. For example, pregnant women and breastfeeding mothers are supposed to receive at least Rs. 6,000 as maternity benefit; children aged between six months and fourteen years should have access to free meals under the midday meal scheme as well via the "Integrated Child Development Services (ICDS)" program in Odisha (Mamata Scheme).

One of the most important aspects of NAFSA is that it can be enforced by law; therefore, individuals have the legal right to demand their food entitlements. This enforceability is crucial, as the demands made on behalf of a State are not just aspirations it should fulfill, but they are supported by mechanisms through which complaints can also be addressed. States are required to establish grievance redressal mechanisms at district and State levels. Most importantly, the NFSA also mandates transparency and accountability measures through the mandatory disclosure of records related to PDS, social audits, and grievance redressal, including the formation of committees, among other provisions. This approach is expected to help combat corruption and inefficiency that have traditionally plagued Global's food distribution systems.

However, the NFSA has introduced its own set of challenges. Misclassification of beneficiaries, pilferage and diversion of food grains, quality concerns on the supply side, and leakages within the delivery system continue to hinder the full achievement of the goals out*lined by the* Act. Additionally, although the NFSA is a crucial step toward ensuring food security, it does not directly address other essential aspects of nutrition security, such as dietary diversity and quality.

Figure 3. Four pillars of food security

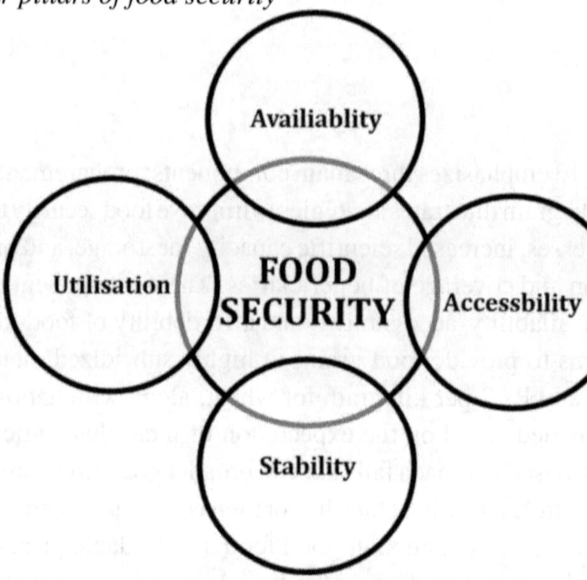

Food security is based on four pillars: Availability, Accessibility, Stability, and Utilization (Guiné et al., 2021). Availability refers to the production and supply of calories, which are put into systems that ultimately lose interest because, one-quarter or two generations down, not enough other foods are available in the broader span. Accessibility draws attention to both economic and physical access, whether individuals have the financial means and the location to acquire nutritious food. While consistent availability and food access are stable, this prevents you from being influenced by economic or environmental shifts. Utilization ensures that the food consumed is safe, nutritious, and sufficient to meet an individual's dietary requirements. All in all, these represent an entire framework for understanding and addressing food security as a whole, which will make both national-level policy and practical interventions for sustainable development more coherent within Global.

5. RESULTS AND DISCUSSION

A. POLICY ANALYSIS

Several important reports have been published by various nodal agencies regarding the right to food and sustainable food security in Global. One such report is the Comptroller and Auditor General of Global Report on implementing the NFSA, 2013, submitted in 2016. It assessed how the NFSA, 2013, was implemented by the States and Union Territories. The report highlighted significant gaps: "By 2015, only 51% of the eligible beneficiaries had been identified, the State governments had failed to make the necessary updating of the existing Ration Cards as per NFSA requirement." It also noted issues related to storage and logistics, as well as delays in supplying food grains to the doorsteps of already registered beneficiaries through the targeted PDS. Additionally, there was a problem with storage, as the States had not taken the necessary actions to provide adequate storage facilities: "physical inspection in the audit of this Office revealed food grain losses in storage ranging between 1.5% and 22.19% with the average loss being 4.11 percent across the test-checked districts." Therefore, it appears reasonable to conclude that there were substantial challenges in implementing the NFSA.

The Shanta Kumar Committee submitted another report on the Long-term Grain Policy, presented to the government in 2014 (on the Food Corporation of Global). The report offers valuable recommendations such as "decentralization of FCI operations and procurement, abolition of levy and non-levy sugar on the PDS scheme, transfer of rail freight on food by the government." Another important suggestion in the report was that maintaining the centralization of food grain procurement is no longer sensible and could lead to a significant waste of public funds. The PDS has experienced considerable "leakage" over the years, and according to the report, the procurement operation should be decentralized.

Figure 4. Maslow's hierarchy of needs theory

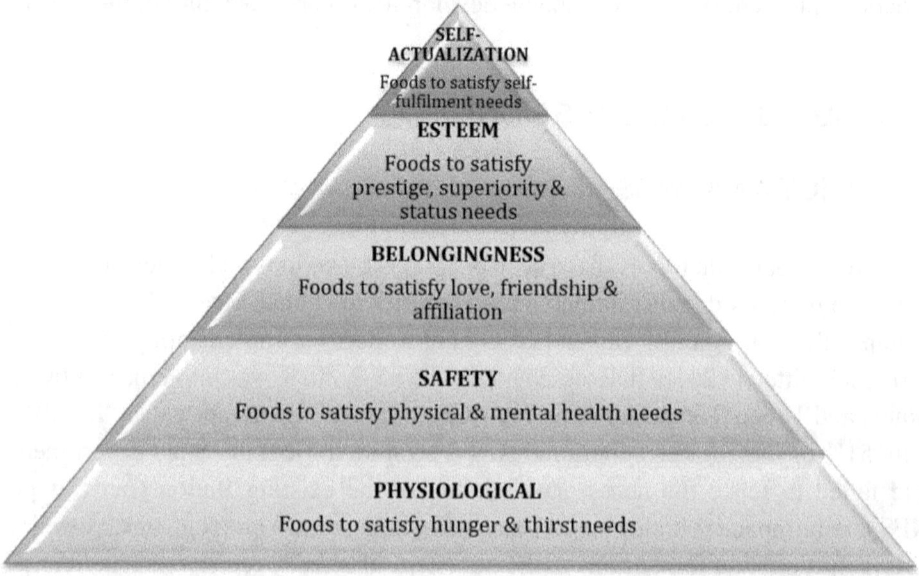

Abraham Maslow introduced the hierarchy of needs theory in his 1943 paper about prevailing dynamic sources. The psychological principles reflect human experiences and are arranged on five levels, with humans unable to reach an optimal level. At the bottom of this pyramid are basic human needs, including food and water. Food is fundamental to physiological requirements and essential for life and health because it supplies the energy and nutrients needed to support bodily functions. Once people satisfy their basic needs, they focus on the secondary ones listed above, such as safety, love, belonging, and esteem (pride in accomplishments), leading to the final stage of self-actualization. By its very nature, food security is the foundation for survival, enabling people and communities to pursue their further social, economic, and psychological needs. Food security serves as the base on which individuals can develop and aspire for better things in life, making it a crucial aspect of human development discussions and policy-making.

B. Existing Policies Impacting Food Security in Global

Food security in Global involves various government policies aimed at ensuring safety, nutrition, and good health, including food production efforts under close supervision from the USDA. The foundation of food security policy has been the NFSA, 2013, which legally guarantees that every family with six or more members in Global has the right to receive grains at subsidized prices, with an upper limit of up to two-thirds of their needs. The provision of food grains under the NFSA was supported by other essential policies, such as the Public Distribution System (PDS), which has been one of the largest distribution networks in the world and primarily aims to deliver entitlements as outlined in Section 3(1), second provision of the NFSA, 2013. These include the PDS, Integrated Child Development Services (ICDS), and Mid-Day Meal Scheme, which address nutritional insecurities for children, pregnant women, and lactating mothers. The programs aim to reduce child malnutrition, improve educational quality, and decrease dropout rates by providing school meals to students. These initiatives are impressive frameworks; however, there is significant variability in Global's decentralized implementation across its broad and diverse geography, highlighting the need for improved governance and accountability. It also examines how initiatives like the "Pradhan Mantri Kisan Samman Nidhi (PM-KISAN) Scheme," which provides income support to small and marginal farmers, will affect food security. This financial help helps farmers stay in business and benefits food safety by contributing to long-term agricultural development. The agrarian credit facilities and crop insurance schemes ensure that farmers are not heavily impacted by economic issues caused by yield uncertainties and climatic challenges, thus maintaining continuous food production.

Nevertheless, deficiencies in storage facilities, mismanagement of the food supply chain, and a corrupt bureaucracy have weakened these policies, and their expected results have not been fully realized. This also causes significant wastage of agricultural produce due to high post-harvest losses caused by poor infrastructure. The situation highlights the need for an immediate and comprehensive review of agricultural policies, modernization of food storage facilities, and technological empowerment in agriculture to increase efficiency while improving logistics and infrastructure development. The importance of technology in strengthening food security cannot be overstated. For example, initiatives like the e-National Agriculture Market (e-NAM) aim to create a seamless and transparent electronic marketplace connecting existing APMCs through an online trading platform. This will make price discovery more efficient, ensuring better returns for farmers. However, both require more extensive extension services and greater integration with traditional supply chains.

Table 1. Evolution of food security policies in global

Policy Name	Year Implemented	Target Population	Key Provisions	Coverage (%)	Reported Outcomes
Public Distribution System (PDS)	Pre-1960	General Population	Subsidized foodgrains	50%	Reduced hunger but issues with accessibility and leakage
Integrated Child Development Services (ICDS)	1975	Children under the age of 6, Pregnant and Lactating Women	Supplementary nutrition, health education	35%	Improved child nutrition and health awareness
Mid-Day Meal Scheme (MDMS)	1995	School going Children	Free lunches in schools	73%	Increased school attendance and reduced child malnutrition
National Food Security Act (NFSA)	2013	67% of the Population	Legal entitlement to subsidized grains, maternity benefits	81%	Enhanced access to food, still assessing long-term impacts

The above table shows Global's ongoing efforts to improve food security through various policies. Each policy builds on earlier ones, introducing more complete measures to ensure food access and nutritional support for different parts of the population.

C. Integrating Sustainability into Food Security Policies

Given the increasing importance of sustainability in global food policy, ignoring issues like those discussed above is not an option. Integrating these priorities is based on the idea that food security should not be achieved at the expense of ecological health or by sacrificing resources for future generations. Global's The urgent task is to embed sustainability principles into all food security policies, especially regarding production practices, supply chain logistics, and consumption patterns, since these factors significantly impact human health and nutrition, along with planetary health.

Sustainable food security policies encompass, among other things, an emphasis on ecologically sustainable farming methods that genuinely benefit the farmers who grow the food at fair rates. The policy might promote traditional practices like crop rotation and biofertilizers that help create healthier soil and, indirectly, support biodiversity by reducing reliance on chemical inputs such as fertilizers and pesticides over time. Here, supporting organic farming corridors (as seen with initiatives from the government of Global in states like Sikkim and Kerala) exemplifies a policy that fosters sustainability and helps conserve local agricultural diversity.

Additionally, sustainable approaches must be developed alongside robust systems for natural resource management that safeguard food security. We need policies to enhance water use efficiency and promote conservation practices since water is the most essential agricultural input. A framework for sustainable agricultural policies includes rainwater harvesting, drip irrigation, and the development of drought-resistant crop varieties. Expanding irrigation with a focus on conservation and management aligns with the goals of "Pradhan Mantri Krishi Sinchayee Yojana (PMKSY)."

Addressing sustainability in food policy also involves rethinking supply chains to reduce the significant and questionable production losses in modern industry. Policies should now promote the development of a truly cold chain, efficient logistics, and market linkages to connect surplus food with end consumers. *Incorpo*rating technology and innovation into the chain, such as blockchain for product tracking or artificial intelligence in logistics, will boost efficiency and transparency, helping to eliminate food waste and ensure consistent availability.

Consumer awareness and education are essential for changing our consumption habits toward sustainability. Nutrition literacy highlights the importance of adjusting dietary practices to be more sustainable. Therefore, policies focused on this area are likely to have the biggest influence on consumer behavior. Labeling programs can display the carbon footprint of food items or promote plant-based diets to encourage more sustainable choices. The "Rashtriya Poshan Abhiyan (National Nutrition Mission)" exemplifies this by aiming to improve nutritional outcomes among children and pregnant or nursing women through behavior change, blending nutrition and sustainable consumption policies. Broader sustainability efforts can be promoted by aligning food security policies with international environmental and health standards. National policies should reflect Global's commitments under the 2030 Agenda for Sustainable Development Goals, especially Goal 2 (Zero Hunger) and Goal 12 (Responsible Consumption and Production), with a focus on sustainable food systems. These initiatives benefit from international agreements that support sustainable lifestyles and offer funding for food security.

Legal rights are considered effective only when there are suitable remedies. This is summarized by the maxim *"Ubi jus ibi remedium,"* meaning where there is a right, there is a remedy. Simply put, this illustrates the connection between a right and a corresponding remedy. However, without strong enforcement mechanisms within the NFSA, the right to food remains purely a formal or theoretical right. The analysis above highlighted significant gaps in policy implementation and enforcement. Therefore, developing robust legal frameworks and ensuring effective enforcement of policies are essential to providing real remedies for those deprived of their right to food.

6. CASE STUDIES

A. Judicial Reinforcement of the Right to Food in Global

The right to life and human dignity are inalienable rights, and the right to food is an essential part of these principles, closely linked to several provisions enshrined in our Constitution. The Constitution of Global, in its Preamble, aims to establish a social order based on justice, liberty, and equality. This core principle inherently encompasses food sovereignty, which is essential for achieving social and economic justice. Therefore, the Preamble upholds values for individual dignity, aligning with the need for food security, which also becomes a fundamental aspect of Global's constitutional governance.

Article 21 of the Constitution, which guarantees the right to life and personal liberty, has been expansively interpreted by the Supreme Court to include a variety of rights such as free legal aid, speedy trial, protection against handcuffing during arrest and detention, etc., under its scope on considerations related to adequate human dignity. This concept and its natural extension to the right to socialism are theoretically connected with a fundamental assumption that an individual cannot live with dignity without physical sustenance. The Directive Principles of State Policy in Articles 39(a) and 47 strengthen the right to food. The State shall, in particular, direct its policies toward securing the rights of citizens (between men and women) equally and providing an adequate means of livelihood.

The case of "Kishen Pattnayak v. State of Orissa (1989)" was a landmark decision where the Court recognized the Globaln State's obligation to provide adequate food and nutrition to its citizens; it has since become a benchmark for all subsequent jurisprudence on the right to food. Another significant case in this context is the "People's Union for Civil Liberties v. Union of Global," commonly known as the "Right to Food Case" (2001). The petition also sought directives to enforce the right to food under Article 21, due to starvation deaths. This case led to the issuance of various interim orders to ensure that government schemes like PDS, ICDS, and MDMS are effectively implemented. Ultimately, this case resulted in the adoption of the NFSA, 2013, which recognized the right to food and made it a legal entitlement, with subsequent orders issued, such as Re 1 per kg rice at BPL prices.

"Kapila Hingorani v. State of Bihar (2003)" reaffirmed the judiciary's commitment to the right to food. In this case, it addressed visual undertrials and prisoners suffering from malnutrition, highlighting the responsibility of the State to ensure the welfare of its residents. The Court's directives in this case reinforced the fundamental principle underlying the Right to Food, namely Article 21, the Right to Life and Personal Liberty. In "Swaraj Abhiyan v. Union of Global (2016)," which dealt with drought management, the Court emphasized that the State must proactively implement relief measures to prevent food security from being compromised during crises, thereby protecting the national government from such blame. The right to food in Global is firmly rooted in constitutional law, with a prominent enforcement mechanism through public interest litigation before the Supreme Court. Through its landmark judgments, the judiciary has consistently reaffirmed that the right to food is an essential component of the right to life, compelling the State to develop comprehensive policies and interventions to eliminate hunger and undernutrition and to ensure that all citizens—regardless of poverty status—have access to adequate food.

B. Best Practices of Sustainable Food Security Models from Global and Worldwide

It has also facilitated the pursuit of sustainable food security through new replication models that combine high-productivity farming with environmental sustainability and socio-economic inclusion. This approach has proven effective in many parts of the world, though different successful models emphasize food security and ecological resilience to varying degrees. By examining these models globally and those developed in Global, this analysis offers insights into their strategies and effects.

Brazil's Zero Hunger program, which combines social assistance with small farmer support and school meal initiatives, serves as a well-known global model. Launched in 2003, the program adopts a comprehensive approach to eradicating hunger and extreme poverty through targeted measures such as food distribution, expanding access to education, and enhancing employment opportunities. Its multifaceted strategy aims to support smallholder farmers by developing local markets and linking food producers directly with consumers, thereby promoting access to commodities at the community level. The Brazilian example shows that government-led policies, community participation, and private-sector involvement can work together to build sustainable food systems.

America and Europe have initiated the Grassroots Food Security model, which is practiced by the local Community Supported Agriculture (CSA) and is considered one of the best practices to be replicable in different parts of the world. In this model, a portion of the farm's produce can be purchased in advance, and locals secure produce and allow farmers to stabilize income, providing some relief from uncertainty around crop production. The model helps develop local economies while promoting organic cultivation and reducing farm-to-market distance, eventually lowering the carbon emission from transportation (Barbosa et al., 2022).

In Global, Kerala exemplifies women's empowerment and community-based food security through its Kudumbashree initiative. Kudumbashree is a network of women's self-help groups (SHGs) dedicated to empowering women as entrepreneurs and farmers. The model organizes SHG women to collaborate and cultivate fallow land collectively, using methods that draw on their traditional knowledge and focus on food production. This socio-economic upliftment of women has boosted both local food production and the business model. The approach demonstrates how women's economic empowerment can improve food security, especially in areas near water bodies, while promoting sustainable development through gender equality. Global has also adopted a new model of ICT-enabled agriculture. ITC Limited's E-Choupal uses digital technology to inform farmers about market prices, weather forecasts, and farming techniques. This has radically transformed the traditional agricultural supply chain into a transparent and efficient ordering system, allowing farmers to make better decisions about crop sales and increasing their returns per ton of potatoes or grain.

Each model offers insights into how sustainable food security practices can be replicated and scaled up. Brazil's zero hunger initiative highlights the importance of political will, governance, and comprehensive policy frameworks. The CSA model is a great example of community engagement and direct market approaches, while the Kudumbashree project in Kerala demonstrates the vital role of women farmers in agriculture and local development. Lastly, E-Choupal exemplifies how technology can transform rural agriculture. A comparison of food security between Global and other countries reveals significant differences in policy effectiveness, implementation challenges, and results. In Global, the very high rates of child stunting and wasting underline serious nutritional problems.

C. Replicability of these Models in Globaln Settings

A wide range of successful sustainable food security models have been implemented worldwide, and the same is true even within specific regions or socio-economic groups. However, whether these models can be replicated in different Globaln settings depends on a more detailed assessment of regional differences in economic factors, cultural practices, and governance structures. Global's geographic and cultural diversity, from the snowy Himalayan mountains to the desert-like regions, shapes its unique climate and topography, resulting in various agricultural practices that rely on different food distribution systems at various levels of industrialization. International food security models, such as the community-supported agriculture (CSA) scheme in the USA or Brazil's zero hunger program, should be carefully evaluated for their relevance to local issues. The CSA model, which primarily relies on community participation and local marketing, could be adapted for the Globaln context by leveraging the existing traditional networks of self-help groups and cooperative societies that are already *effectively operating in rural Global*.

Additionally, learning from Brazil's success with its zero hunger approach, cash transfer scheme, and food programs is the right policy to facilitate entitlement for many hungry people (*Assessing Multidimensional Sustainability: Lessons from Brazil's Social Protection Programs - PMC*, 2020). Nevertheless, these adaptations must also target specific challenges, such as land ownership patterns and water stress, and improve access to technology. To get around this, programs must be localized and embedded in strong policy frameworks that reward sustainable farming methods and promote equity along market chains. While looking at various feasibility options for replication of best practices in Globaln settings, technology plays a critical role in reproducing the food security models that have been successful. Global is perhaps a poster child for digital infrastructure and has huge potential to use technology more seriously with better crop yield forecasting, precision agriculture practices, and reduced food wastage in supply chains, thereby improving the distribution of food within Global (Yashvardhan et al., 2022). For instance, real-time weather support can be provided to farmers via digital lookalikes or access information on market prices and government schemes through service providers using the same TPM solutions (*How Can The 'Digital Global' Programme Help Farmers To Improve Farm Productivity And Income?*, 2024)

A. Gaps within the current Policy Frameworks

Globaln legislation, including the NFSA, 2013, and associated policies, has arguably established one of the most comprehensive legal frameworks for securing the right to food in Global. However, many gaps within existing policy structures continue to hinder the realization of human rights. These gaps manifest as a disconnect between what legislators expect their law to achieve and how food security actually functions. The first issue is that, although the scope of the NFSA is broad, it overlooks many factors that determine food insecurity—the numerous aspects related to access to food, nutritional quality, and safety. The Act mainly addresses the calorie content of food, which is only a small part of fighting malnutrition, stunting, and wasting among children. This oversight leads to a situation where prioritizing increased food quantities takes precedence over considerations of its quality, allowing us to celebrate food security without effectively tackling malnutrition.

Second, the NFSA targets poor households by relying on periodically updated but outdated census data; this has caused many genuinely needy populations to be excluded because of incorrect or old information. As a result, numerous potential beneficiaries are left out of the Act's coverage. Critics often apply eligibility criteria too narrowly, limiting aid programs' ability to reach a larger needy population. The public distribution system (PDS), which is the main way food entitlement is delivered under NFSA, also faces issues with leaks and corruption, leading to significant access challenges. The NFSA operates independently, not integrated with even a basic level program, nor linked to other crucial sectors such as agriculture, health, and sanitation, which are essential for a comprehensive food security strategy. This disjointed approach makes these efforts less effective because the problems are interconnected. Food security is also related, and similar governance issues can arise from agricultural policies that conflict with nutrition goals-such as prioritizing staples over more nutritious crops, ultimately contributing to widespread undernutrition.

Meanwhile, efforts to legally enforce the right to food remain highly controversial. Although the Supreme Court of Global has been proactive on right-to-food issues, much of the enforcement burden falls on a judiciary already overwhelmed with a heavy workload. The legislative and executive branches do not consistently follow up to ensure compliance with court orders for food supply. This gap between judicial directives and executive action weakens the effectiveness of existing legal frameworks.

Figure 5. Challenges in NFSA

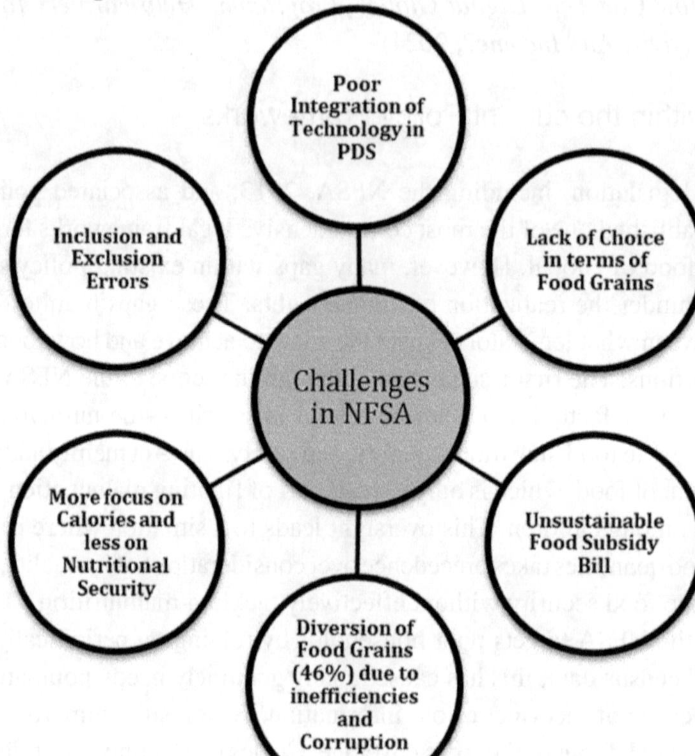

There is a vast, unbridgeable gap in the awareness and education among citizens about their entitlements under NFSA. Large numbers of people are either unaware of their entitlements or the processes they need to follow if there is even an appearance that those responsible for enforcing compliance did not perform as mandated. This lack of accountability measures infringes upon the empowerment needed to implement this legislation effectively.

B. Challenges in Implementing the Right to Food and Sustainable Practices

Several challenges on the ground stem from systemic and operational barriers that hinder the effective realization of the right to food and sustainable practices within Global. This consistently highlights the difficulty of translating policy into practice, even with strong legal frameworks such as programs like the NFSA, 2013. The logistical challenge of ensuring food reaches everyone fairly across Global's vast and diverse geography begins with rural areas, where poor infrastructure can prevent food from reaching people. This vulnerability is then worsened by the seasonality of events like monsoons and other natural disasters, which disrupt supply chains with alarming regularity. These geographic and infrastructural shortfalls greatly limit the state's ability to provide a consistent food supply, thereby hampering the practical realization of the right to food.

Additionally, governance and coordination among the many stakeholders responsible for implementing food security policies add another layer of complexity. The Globaln food security system is a mix of different feeding programs; some focus on supplementary nutrition to prevent malnutrition or provide emergency supplies and services during economic crises or natural disasters, while others aim to tackle chronic hunger and malnutrition by providing staple foods. Different states have various food security mechanisms, allowing them to achieve different levels of hunger alleviation, as explained earlier. It encompasses both long-term welfare programs, emergency responses, and short-term interventions whenever needed. However, poor coordination and unclear accountability structures lead to inefficiencies and corruption, undermining the objectives of the NFSA. Common issues such as pilferage, inefficient food management practices, and bureaucratic delays further damage trust in an already fragile delivery *system.*

The second major hurdle is that, because of economic reasons, the scope of activities aimed at ensuring food security may be limited. When food programs are not properly funded or budgeted to reflect their actual costs, they underperform and ultimately fail to achieve their full purpose. Economic constraints also limit the quality and nutritional value of food provided through subsidy programs, raising questions about whether such initiatives meet the dietary needs of the vulnerable populations they serve. Socio-cultural factors are also significant in creating implementation barriers, in addition to economic and logistical issues. Food aid is often distributed based on discrimination related to caste, religion, and gender, causing marginalized groups in Global to face inequality in food resources. Farming and agriculture have borne the impact of such deep-rooted biases, making it essential to adopt a more inclusive approach in policy development to ensure that food security measures are a*ccessible to all equally*.

In addition, the enshrinement of sustainable practices in bringing about the right to food presents further layers. The provision of food is not just sufficient to achieve sustainable food security, it should cover the source, production process, and consumption per se (*Assessment and Principles of Environmentally Sustainable Food and Agriculture Systems*, 2023). A change is needed to sustainable farming practices, a new paradigm that many of the traditional stakeholders in agriculture are reluctant to embrace. Obstacles, including the lack of public understanding about sustainable agriculture practices and underprepared education systems on them as well as resistance to change from stakeholders in chemical-dependent agricultural industries, seem to be impeding advances (Priya & Singh, 2022).

7. SUSTAINABILITY IN FOOD SECURITY

A. Integrating sustainable practices within the right to food

The right to food is essential for the sustainable development of people, regardless of their country of origin. The sustainability of agricultural and food systems is crucial for providing safe, nutritious, and sustainable food to meet the needs of a growing global population without threatening air, water, or Earth's resources. A multifaceted approach is needed, including adopting sustainable farming practices, reducing food waste, and promoting fair distribution of food. Transitioning to sustainable agricultural practices involves shifting from traditional, intensive farming methods to more sustainable systems, often organic and agroecological. Key features of these practices include improving soil health and reducing chemical inputs like fertilizers and pesticides, which enhances crops' resilience to climate stresses. Techniques such as crop rotation, intercropping, and natural pest control not only lessen environmental impacts but also contribute to better nutritional quality of food. Additionally, sustainable agriculture helps smallholder farmers become more profitable and resilient while ensuring food security both locally and nationally.

This is another important aspect of sustainability related to the right to food. Globally, one-third of all food produced is lost or wasted—resources go to waste and meals go unclaimed by those who need them. Strategies to reduce food waste include better food storage and transportation methods, improved food production practices (e.g., increasing crop yields), removing urban barriers to the sale of "less than perfect" fruits and vegetables, and updating expiration date labeling regulations to align with the federal Food Date Labeling Act of 2019, which calls for clear and consistent placement. A single label on all products indicating freshness or spoilage can provide economic incentives to lower surplus accumulation of whole fruits and vegetables, while preventing oversized portions from being displayed under highly filtered circumstances that can harm consumer satisfaction. Cash bonus payouts are regularly offered to motivate different purchasing behaviors, and these incentives can help promote more normalized and responsible consumption.

Similarly, the fair distribution of food is essential to help us sustain ourselves, especially regarding sustainable access, and to make what we are calling here 'integrous' a new term created for this principle. This means every segment of the population can access a nutritious diet, regardless of their socio-economic status. Achieving this can involve, among other approaches, designing market systems that expand small producers' access to markets, implementing social protection programs to increase their purchasing power (including their ability to access food), and establishing backyard production nodes closer to metropolitan centers, thereby reducing transportation costs and related issues. Additionally, the legal authority to allocate natural resources within communities is vital for ongoing food activities. This includes preventing the displacement of indigenous and local communities from their ancestral lands, where they sustainably manage resources, and safeguarding traditional knowledge associated with farming practices that are often inherently sustainable.

This means that sustainability is a legal obligation for exercising the right to food and can only be achieved through policy environments that promote incentives, research, education, and sustainable agricultural models for farmers transitioning from traditional farming. Current policies are unsustainable, requiring collaborative efforts from governments, international organizations, and NGOs to support sustainable food systems. This includes adding sustainability criteria to agricultural subsidy frameworks, investing in research for sustainable farming technologies, and hosting international platforms for sharing best practices.

The diagram above illustrates the complex relationships that connect food security, food crises, and financial crises. Food insecurity can trigger financial crises, which can then impact financial and economic stability. This can further worsen food insecurity, creating a cycle where financial crises feed into ongoing food insecurity. Understanding these dynamics is essential for developing policies that sustainably address long-term food insecurity by targeting its root causes in the economy and politics. This highlights the importance of a policy environment that considers the interconnectedness of food insecurity with other aspects of well-being, such as financial security.

B. Role of climate change and environmental sustainability in shaping food security strategies

A better understanding of the relationship between climate change and food security is now seen as a key part of sustainable development strategies. Climate change influences many aspects of food production systems, leading to less availability, access, and reliability of the global food supply. The rising frequency and intensity of extreme weather events, such as severe droughts, floods, and hurricanes, which come with increasing global temperatures, have become threats to agricultural productivity. As a result, these extreme conditions can damage crop yields and harm livestock systems and fisheries that are vital to global food security.

In South Asia and Sub-Saharan Africa, regions that are already highly vulnerable to the effects of climate change, these impacts have been more intense, further increasing vulnerabilities to food security. For example, changes in rainfall affect rainfed and irrigated agriculture; sea level rise and salinization can make fertile land infertile. Given such environmental shifts, some farming practices must be reconsidered, and climate resilience must be integrated into agricultural responses for food security. Therefore, agricultural adaptation strategies are crucial for addressing climate change impacts while supporting sustainable food production systems. These strategies include developing and promoting climate-resilient crop varieties, advancing agricultural technologies, and improving water management techniques. However, agroecological methods such as crop diversification, conservation agriculture practices, and sustainable land management also provide valuable options for reducing climate change effects across landscapes. Furthermore, policy frameworks play a vital role in tackling climate change challenges and ensuring food security. Practical and effective policies are needed to promote sustainable agricultural practices that increase food productivity and reduce environmental footprints. These include policies supporting sustainable land use, water conservation, and emissions reduction. Building a strong global food system also requires international cooperation through sharing information, exchanging technology, and providing resources. Integrating climate change adaptation into national food security plans and policies is essential. Policymakers should prioritize food security in climate change action plans and invest in sustainable agricultural practices to help food production and distribution adapt to changing climatic conditions. This also involves investing in sustainable agriculture research and development to address the challenges posed by our changing climate. Governments, the private sector, civil society, and the international community will share responsibility for safeguarding food security in a world destabilized by climate change. A comprehensive solution must consider the short-term effects of climate disruptions and emphasize resilience and sustainability for future global food security. Recognizing the importance of environmental sustainability in establishing food security, strategies should support this collective effort and lead us toward a more climate-resilient future where hunger can be prevented.

8. RECOMMENDATIONS FOR POLICY ENHANCEMENTS

A thorough reassessment of policy frameworks is essential to tackle the challenges of increasing food insecurity and environmental damage. They propose several policy reforms to enhance sustainability and reinforce the legal right to food. It is vital to concentrate on the specific legal frameworks introduced to address these issues and to improve the effectiveness of food security policies. The *principle of "Lex specialist derogate legi generali"* can help establish this. In fact, the NFSA can be viewed as a law specialized in food security, meaning general measures are replaced and do not need to be implemented. Thus, it becomes possible to better customize the strategies to meet particular needs and refine the approach within the legal system.

A. Proposals for Policy Amendments to Enhance Sustainability

First, sustainability guidelines must be embedded in all agricultural policies to prevent long-term food production from damaging ecological functions. Policies should support and promote sustainable agricultural practices, such as permaculture, organic farming, and conservation agroforestry, to maintain healthy soils and prevent erosion. The potential for carbon sequestration of these methods can be increased by up to 5% compared to traditional agriculture. Farmers should adopt these sustainable techniques, and to motivate this, existing subsidy programs and financial incentives need to be revised and updated. Water management policies also need an urgent overhaul to improve the efficiency of crop irrigation. More effec*tive* Irrigation methods, such as drip irrigation and rainwater harvesting, should be implemented immediately. Furthermore, policies should motivate farmers to recycle water used in agriculture and to use treated wastewater, helping to lessen pressure on the limited freshwater resource.

B. Recommendations for Strengthening the Legal Framework to Support the Right to Food

Embedding food security and sustainability into national constitutions and legislative instruments is essential for strengthening a rights-based framework. It should explicitly recognize the right to food and outline how the State will meet its obligations in this area, ensuring that all actions and policies focus on this goal on this goal. Therefore, we must ensure that our legal frameworks include systems of accountability and enforcement. This involves establishing independent bodies or ombudspersons responsible for monitoring, reviewing, and enforcing commitments related to food security. It should also handle complaints about violations of the right to food, transforming the right into a principle and an actionable legal benefit. Ensuring that food security impact assessments become mandatory for all development projects at both the planning and implementation stages, through appropriate policy development, is crucial. These assessments help identify and mitigate potential negative impacts on food security before project approval. They must be transparent and include the community perspective*s* to reflect local insights and concerns. To address the transboundary nature of food security, national laws should align with international agreements and protocols related to food, the environment, and trade. This alignment enables more coherent and effective management of transboundary issues, such as food imports and exports, shared water resources, and genetic resources for food and agriculture.

9. CONCLUSION

This study highlights the connections between policy effectiveness, legal measures, and public sustainable food security. NFSA, 2013, marks an important step in enshrining the right to food. However, its implementation faces ongoing challenges, such as systemic inefficiencies, misclassification of beneficiaries, leakages in the Public Distribution System (PDS), and governance issues. Additionally, the link between food security and climate change was inadequately addressed, underscoring the need for integrated efforts around sustainability and resilience. While judicial interventions have helped strengthen the right to food, they often lack consistent institutional enforcement, leading to varied food security outcomes across regions. Although the research offers a broad overview with high-level analysis, it has limitations. Primarily, it is doctrinal, relying on policies, legal texts, and secondary data. While this provides valuable theoretical insights, it lacks field-level empirical data, such as experiences from beneficiaries or lessons from government officials. Incorporating case studies and interviews through field-based qualitative and quantitative methods could address this gap in future research. Second, the study centers on national policies and laws, with limited comparative analysis of best practices in global food security governance. Broader cross-country comparisons could offer better insights into effective models Global might adopt. Third, the research does not extensively cover the economic aspects of food security, including market forces, agricultural pricing policies, and trade dynamics, which play a crucial role in food access. These areas require further exploration within a comprehensive policy framework. Achieving collective action will depend on legal enforcement, policy coherence, and developing climate-resilient farming systems to expand food security in Global. The importance of aligning sustainability, governance reforms, and community-based initiatives to ensure equitable food access highlights the need for future research to empirically assess policy effectiveness, examine global policy models, and analyze economic drivers to build a more holistic framework for a food-secure Global.

ACKNOWLEDGEMENT

The authors are indebted to KIIT University, Bhubaneswar, for extending academic support and research facilities. They would like to express sincere gratitude to the editorial and peer-review boards for their constructive suggestions, which helped improve the quality of the paper.

REFERENCES

Barbosa, S. B., de Melo, G. A., Peixoto, M. G. M., Mendonça, M. C. A., & de Andrade Guerra, J. B. S. O. (2022). Community Supported Agriculture (CSA) to promote Food Security and Sustainable Agriculture. In Leal Filho, W., Kovaleva, M., & Popkova, E. (Eds.), *Sustainable Agriculture and Food Security*. World Sustainability Series. Springer., DOI: 10.1007/978-3-030-98617-9_26

FAO. IFAD, UNICEF, WFP and WHO. (2023). The State of Food Security and Nutrition in the World 2023. Urbanization, agrifood systems transformation and healthy diets across the rural–urban continuum. Rome, FAO. https://doi.org/DOI: 10.4060/cc3017en

Food and Agriculture Organization of the United Nations. Climate Smart Agriculture Sourcebook, Sustainability. *Food Security and Climate Change: Three Intertwined Challenges*. https://www.fao.org/climate-smart-agriculture-sourcebook/concept/module-a1-introducing-csa/chapter-a1-1/en/

Global Hunger Index. (2024). 2023 Global Hunger Index: The Power of Youth in Shaping Food Systems-Africa Brief.

Guiné, R. de P. F., Pato, M. L. de J., da Costa, C. A., da Costa, D. de V. T. A., da Silva, P. B. C., & Martinho, V. J. P. D. (2021). Food Security and Sustainability: Discussing the Four Pillars to Encompass Other Dimensions. *Foods*, *10*(11), 2732. DOI: 10.3390/foods10112732 PMID: 34829013

National Food Security Portal, Department of Food and Public Distribution Division, Government of Global. *National Food Security Act (NFSA) 2013*. https://nfsa.gov.in/portal/nfsa-act

Priya, P., & Singh, S. (2022). Factors Influencing the Adoption of Sustainable Agricultural Practices: A Systematic Literature Review and Lessons Learned for Global. *The Forum for Social Economics*, *53*, 1–17. DOI: 10.1080/07360932.2022.2057566

Sen, A. (1983). *Poverty and Famine: An Essay on Entitlement and Deprivation.* Oxford University Press., https://academic.oup.com/book/32827

Supreme Court of Global. *Kishen Pattnayak and Anr. Vs State of Orissa on 9 January, 1989.* https://globalnkanoon.org/doc/181576930/

Supreme Court of Global. *Kapila Hingorani vs State of Bihar on 9 May 2003.* https:// globalnkanoon.org/doc/1455798/

United Nations. Human Rights, Office of the High Commissioner (1966). General Assembly Resolution 2200A (XXI). *International Covenant on Economic, Social and Cultural Rights.* https://www.ohchr.org/en/instruments-mechanisms/instruments/ international-covenant-economic-social-and-cultural-rights

United Nations. Meeting Coverage and Press Releases (13 February 2024). Climate Action Can Help Fight Hunger, Avoid Conflicts, Official Tells Security Council, Urging Greater Investment in Adaptation, Resilience, Clean Energy. https://press .un.org/en/2024/sc15589.doc.htm

United Nations. Human Rights, *The Right to Adequate Food, Fact Sheet No. 34.* https://www.ohchr.org/sites/default/files/Documents/Publications/FactSheet34en .pdf

United States Environmental Protection Agency US-EPA. (2025). Climate Impacts on Agriculture and Food Supply, Climate Change Impacts. https://www.epa.gov/ climateimpacts/climate-change-impacts-agriculture-and-food-supply

Yashvardhan, V., Pyare, R., & Singh, R. (2022). Technological Advancement in Digital Farming and Global's Present Scenario: A Review. *International Journal of Environmental & Agriculture Research, 8*(12). Advance online publication. DOI: 10.5281/zenodo.7496469

Compilation of References

Abdalla, R. (2016). *Introduction to geospatial information and communication technology (GeoICT)*. Springer.

Abdou, A. H., Mohamed, S. A. K., Khalil, A. A. F., Albakhit, A. I., & Alarjani, A. J. N. (2022). Modeling the relationship between perceived service quality, tourist satisfaction, and tourists' behavioral intentions amid COVID-19 pandemic: Evidence of yoga tourists' perspectives. *Frontiers in Psychology*, *13*, 1003650. PMID: 36211929

Abu Bakar, M. R. (2025). A systematic review of smart campus initiatives in transforming higher education institutions. [JIKM]. *Journal of Information and Knowledge Management*, *15*(SI2), 176–188.

Acharya, S. S., Sen, S., Punia, M., & Reddy, S. (Eds.). (2017). *Marginalization in globalizing Delhi: issues of land, livelihoods and health* (pp. xxxix+-465). Springer India.

Adiba, M. A. (2023). *Cause of repetitive fire disaster in informal settlements and the importance of community based disaster risk mitigation* (Doctoral dissertation, Brac University).

Agoo, J., Lanuza, R. J., Lee, J., Rivera, P. A., Velasco, N. O., Guillermo, M., & Fernando, A. (2024). Geographic information system-based framework for sustainable small and medium-sized enterprise logistics operations. *ISPRS International Journal of Geo-Information*, *14*(1), 1.

Ahl, A., Goto, M., & Yarime, M. (2020). Smart technology applications in the woody biomass supply chain: Interview insights and potential in Japan. *Sustainability Science*, *15*(5), 1531–1553.

Ahmada, I. (2024). Advanced sensors, monitoring. Smart Cities and Sustainable Manufacturing: Innovations for a Greener Future, 47.

Ajani, O. A., Gamede, B. T., & Bansal, R. (2024). Challenges and opportunities of integrating industry 5.0 and emerging technologies in higher education for enhancing employability skills. *Industry 5.0 and Emerging Technologies: Transformation Through. Technology and Innovation*, •••, 217–230.

Akgün, A. A., Baycan, T., & Nijkamp, P. (2015). Rethinking on sustainable rural development. *European Planning Studies*, *23*(4), 678–692.

Akindele, O., Ajayi, S., Oyegoke, A. S., Alaka, H. A., & Omotayo, T. (2025). Application of Geographic Information System (GIS) in construction: A systematic review. *Smart and Sustainable Built Environment*, *14*(1), 210–236.

Aladaileh, M. J., Lahuerta-Otero, E., & Aladayleh, K. J. (2024). Mapping sustainable supply chain innovation: A comprehensive bibliometric analysis. *Heliyon*, *10*(7). PMID: 38623205

Ali, A., Imran, M., Jabeen, M., Ali, Z., & Mahmood, S. A. (2023). Factors influencing integrated information management: Spatial data infrastructure in Pakistan. *Information Development*, *39*(2), 213–234.

Ali, A., & Mohammed, S. (2021). Using GIS-based multicriteria analysis for renewable energy site selection: A review. *Renewable & Sustainable Energy Reviews*, *143*, 110–120.

Allaf Jafari, E., Asayesh, F., & Rostamian, M. (2025). Towards building a sustainable future with the tech revolution. *International Journal of Human Capital in Urban Management*, *10*(2), 317–328.

Altarez, R. D. D., Apan, A., & Maraseni, T. (2024). The perspectives of remote sensing and GIS on military environmental impacts: A systematic review. *Environmental Monitoring and Assessment*, *197*(1), 113. PMID: 39738966

Alzahrani, N. A., Abdullah, S. N. H. S., Adnan, N., Ariffin, K. A. Z., Mukred, M., Mohamed, I., & Wahab, S. (2024). Geographic information systems adoption model: A partial least square-structural equation modeling analysis approach. *Heliyon*, *10*(15). PMID: 39170420

Asgharizadeh, E., Jooybar, S., Mahdiraji, H. A., & Garza-Reyes, J. A. (2022). A novel travel time estimation model for modeling a green time-dependent vehicle routing problem in food supply chain. *Sustainability*, *14*(14), 8633.

Ashley, C., & Maxwell, S. (2001). Rethinking rural development. *Development Policy Review*, *19*(4), 395–425.

Bansa, R., Bansal, T., & Pruthi, N. (2024). Demystifying the Benefits and Challenges of Virtual Learning for Differently Abled Students. In *Virtual Lifelong Learning: Educating Society with Modern Communication Technologies* (pp. 70-84). Bentham Science Publishers.

Bansal, R., Martinho, C., Pruthi, N., & Aggarwal, D. (2024). From virtual observations to business insights: A bibliometric review of netnography in business research. *Heliyon, 10*(1). PMID: 38163120

Bansal, R., Pruthi, N., Bansal, T., & Chaudhary, A. (2023). Diving into the consumer's mind: Tools, roles, and ethical concerns of neuromarketing. In *Applications of Neuromarketing in the Metaverse* (pp. 206–222). IGI Global.

Bansal, R., Saini, S., Ngah, A. H., & Durga Prasad, T. (2024). Proselytizing the potential of influencer marketing via artificial intelligence: Mapping the research trends through bibliometric analysis. *Cogent Business & Management, 11*(1), 2372889.

Bansal, R., Shukla, G., Gupta, A., Singh, A., & Pruthi, N. (2023). Optimizing Augmented Reality and Virtual Reality for Customer Engagement. In *Promoting Consumer Engagement Through Emotional Branding and Sensory Marketing* (pp. 24–35). IGI Global.

Bansal, V. K. (2020). Use of GIS to consider spatial aspects in construction planning process. *International Journal of Construction Management, 20*(3), 207–222.

Barbosa, S. B., de Melo, G. A., Peixoto, M. G. M., Mendonça, M. C. A., & de Andrade Guerra, J. B. S. O. (2022). Community Supported Agriculture (CSA) to promote Food Security and Sustainable Agriculture. In Leal Filho, W., Kovaleva, M., & Popkova, E. (Eds.), *Sustainable Agriculture and Food Security*. World Sustainability Series. Springer., DOI: 10.1007/978-3-030-98617-9_26

Basso, B., & Antle, J. (2020). Digital agriculture to design sustainable agricultural systems. *Nature Sustainability, 3*(4), 254–256.

Batty, M. (2018). *Inventing future cities*. The MIT Press.

Beaunoyer, E., Dupéré, S., & Guitton, M. J. (2020). COVID-19 and digital inequalities: Reciprocal impacts and mitigation strategies. *Computers in Human Behavior, 111*, 106424. PMID: 32398890

Bennett, R., & Joyce, J. (2019). *GIS for sustainable development*. Routledge.

Bertolini, M., De Matteis, G., & Nava, A. (2024). Sustainable last-mile logistics in economics studies: A systematic literature review. *Sustainability, 16*(3), 1205.

Bhatta, B. (2012). *Urban growth analysis and remote sensing: a case study of Kolkata, India 1980–2010*. Springer Science & Business Media.

Bhatta, B., Saraswati, S., & Bandyopadhyay, D. (2010). Urban sprawl measurement from remote sensing data. *Applied Geography (Sevenoaks, England)*, *30*(4), 731–740.

Bill, R., Blankenbach, J., Breunig, M., Haunert, J. H., Heipke, C., Herle, S., & Werner, M. (2022). Geospatial information research: State of the art, case studies and future perspectives. *PFG–Journal of Photogrammetry. Remote Sensing and Geoinformation Science*, *90*(4), 349–389. PMID: 36158628

Blasch, E. P., Pellechia, M., Deignan, P. B., Palaniappan, K., Dockstader, S. L., & Seetharaman, G. (2011, July). Contemporary concerns in geographical/geospatial information systems (GIS) processing. In *Proceedings of the 2011 IEEE National Aerospace and Electronics Conference (NAECON)* (pp. 183-190). IEEE.

Boers, B., & Cottrell, S. (2007). Sustainable tourism planning and development. *Tourism Geographies*, *9*(4), 389–407.

Bridgelall, R. (2023). Unlocking drone potential in the pharma supply chain: A hybrid machine learning and gis approach. *Standards*, *3*(3), 283–296.

Brown, G., & Kyttä, M. (2018). Key issues and research priorities for public participation GIS (PPGIS): A synthesis based on empirical research. *Applied Geography (Sevenoaks, England)*, *95*, 123–136.

Bruckmeier, K., & Tovey, H. (2008). Knowledge in sustainable rural development: From forms of knowledge to knowledge processes. *Sociologia Ruralis*, *48*(3), 313–329.

Bui, T. D., Tsai, F. M., Tseng, M. L., Tan, R. R., Yu, K. D. S., & Lim, M. K. (2021). Sustainable supply chain management towards disruption and organizational ambidexterity: A data driven analysis. *Sustainable Production and Consumption*, *26*, 373–410. PMID: 33015266

Carney, S. (2022). Reimagining our futures together: a new social contract for education: by UNESCO, Paris, UNESCO, 2021, 186 pages, ISBN 978-92-3-100478-0.

Ceccato, V. A., & Snickars, F. (2000). Adapting GIS technology to the needs of local planning. *Environment and Planning. B, Planning & Design*, *27*(6), 923–937.

Census of India. (2001). *Primary Census Abstract, Census of India*. Gov't. of India.

Chambers, R. (2006). Participatory mapping and geographic information systems: Whose map? Who is empowered? *The Electronic Journal on Information Systems in Developing Countries*, *25*(1), 1–11.

Chambers, R., & Conway, G. (1992). *Sustainable rural livelihoods: Practical concepts for the 21st century*. Institute of Development Studies.

Changmai, S. (2023). Geospatial Application for Dairy Supply Chain Management. *Journal of Genomics*, *17*(2), 174–183.

Chanu, C. S., Elango, L., & Shankar, G. R. (2021). A geospatial approach for assessing the relation between changing land use/land cover and environmental parameters including land surface temperature of Chennai metropolitan city, India. *Arabian Journal of Geosciences*, *14*(2), 132.

Chatterjee, U., & Majumdar, S. (2022). Impact of land use change and rapid urbanization on urban heat island in Kolkata city: A remote sensing based perspective. *Journal of urban. Management*, *11*(1), 59–71.

Chen, E., Zhou, Z., Li, R., Chang, Z., & Shi, J. (2024). The multi-fleet delivery problem combined with trucks, tricycles, and drones for last-mile logistics efficiency requirements under multiple budget constraints. *Transportation Research Part E, Logistics and Transportation Review*, *187*, 103573.

Cheng, K., Yin, J., Wang, F., & Wang, M. (2025). The impact pathway of new quality productive forces on regional green technology innovation: A spatial mediation effect based on intellectual property protection. *PLoS One*, *20*(4), e0319838. PMID: 40198691

Chen, S., Liu, H., Chen, J., Li, J., Jiang, N., & Ho, H. C. (2025). Community capital, gender, and depression in older caregivers for people living with dementia: A moderated mediational analysis with urban big geospatial data. *Environmental and Sustainability Indicators*, •••, 100785.

Chen, Y., Huang, Z., Ai, H., Guo, X., & Luo, F. (2021). The impact of GIS/GPS network information systems on the logistics distribution cost of tobacco enterprises. *Transportation Research Part E, Logistics and Transportation Review*, *149*, 102299.

Chettry, V. (2022). Geospatial measurement of urban sprawl using multi-temporal datasets from 1991 to 2021: Case studies of four Indian medium-sized cities. *Environmental Monitoring and Assessment*, *194*(12), 860.

Cooke, P. (2001). Regional innovation systems, clusters, and the knowledge economy. *Industrial and Corporate Change*, *10*(4), 945–974.

Coops, N. C., & Wulder, M. A. (2019). Breaking the habit: Remote sensing in ecosystem monitoring. *Ecological Indicators*, *98*, 1–5.

Dangermond, J., & Goodchild, M. F. (2020). Building geospatial infrastructure. *Geo-Spatial Information Science*, *23*(1), 1–9.

Das, S., & Angadi, D. P. (2022). Land use land cover change detection and monitoring of urban growth using remote sensing and GIS techniques: A micro-level study. *GeoJournal*, *87*(3), 2101–2123.

Das, S., Rao, N. V. J., Mishra, D., & Bansal, R. (2024). A review of the pre- and post-COVID-19 effects on the tourism and entertainment industries: Innovative methods and predicted challenges. In *Utilizing Smart Technology and AI in Hybrid Tourism and Hospitality* (pp. 98–117). IGI Global.

Davis, J., Pijawka, K. D., Wentz, E., Hale, M., & King, D. A. (2021). Evaluating geodesign for community-based tribal planning: The role of planners in marginalized communities. *Journal of the American Planning Association*, *87*(4), 527–541.

De Sousa Saldanha, E., & da Silva, H. B. (2024). Linking Tourists' Motivation to Behavior Intentions through Government Policies and Tourists' Satisfaction: Evidence from Ramelau Mountain as a Tourist Destination in Timor-Leste. *Timor Leste Journal of Business and Management*, *6*, 12–25.

Dede, C. (2009). Immersive interfaces for engagement and learning. *Science*, *323*(5910), 66–69. PMID: 19119219

Delgado, F., Garrido, S., & Bezerra, B. S. (2025). Barriers to Visibility in Supply Chains: Challenges and Opportunities of Artificial Intelligence Driven by Industry 4.0 Technologies. *Sustainability*, *17*(7), 2998.

Deng, Y., Gan, V. J., Das, M., Cheng, J. C., & Anumba, C. (2019). Integrating 4D BIM and GIS for construction supply chain management. *Journal of Construction Engineering and Management*, *145*(4), 04019016.

Deng, Y., Wang, X., & Ma, C. (2024). The impact of sports tourism motivation on tourist loyalty: The chain mediation effect of experience quality and tourist satisfaction. *International Journal of Tourism Research*, *26*(5), e2769.

Desalegn, T., Cruz, F., Kindu, M., Turrión, M. B., & Gonzalo, J. (2014). Land-use/land-cover (LULC) change and socioeconomic conditions of local community in the central highlands of Ethiopia. *International Journal of Sustainable Development and World Ecology*, *21*(5), 406–413.

Dhanaraj, K., & Angadi, D. P. (2022). Land use land cover mapping and monitoring urban growth using remote sensing and GIS techniques in Mangaluru, India. *GeoJournal*, *87*(2), 1133–1159.

Dhanushya, J., Ruba, M., Ponkarthika, M., Soundarya, B., Sakthi Priya, V., Dhanushri, S., & Harikrishnan, S. (2025, April). AI-Powered Multipurpose Agrobot for Precision Farming: Crop Health Management, Fertilizer Recommendation and Yield Prediction. In *2025 International Conference on Recent Advances in Electrical, Electronics, Ubiquitous Communication, and Computational Intelligence (RAEEUCCI)* (pp. 1-6). IEEE.

Ding, X., & Xu, J. (2024). Urban professionals' restorative tourism: Exploring the role of perceived environmental restorativeness, push and pull motivations and destination attributes on tourism expectations. *Frontiers in Psychology, 15*, 1293050. PMID: 38566934

Dixit, A., Routroy, S., & Dubey, S. K. (2022). Analyzing the operational barriers of government-supported healthcare supply chain. *International Journal of Productivity and Performance Management, 71*(8), 3766–3791.

Dutta, I., & Das, A. (2019). Exploring the dynamics of spatial inequality through the development of sub-city typologies in English Bazar Urban Agglomeration and its peri urban areas. *GeoJournal, 84*(4), 829–849.

Dwivedi, S., Singh, A. K., Gupta, A. K., & Chand, S. (2024). Introduction to Sustainable Development and Geospatial Technologies. In *Sustainable Development and Geospatial Technology: Volume 1: Foundations and Innovations* (pp. 1-16). Cham: Springer Nature Switzerland.

Ekeh, A. H., Apeh, C. E., Odionu, C. S., & Austin-Gabriel, B. (2025). Leveraging machine learning for environmental policy innovation: Advances in Data Analytics to address urban and ecological challenges. *Gulf Journal of Advance Business Research, 3*(2), 456–482.

El Gareh, F., Elmenssouri, A., Oulamine, A., & Hussain, Z. (2025). A PRISMA-Based Systematic Review on Organizational Commitment and Logistic Performance. *Knowledge Sharing and Fostering Collaborative Business Culture*, 213-240.

Engeström, Y. (2014). *Learning by expanding: An activity-theoretical approach to developmental research.* Cambridge University Press.

Esfahbodi, A., Zhang, Y., Liu, Y., & Geng, D. (2023). The fallacy of profitable green supply chains: The role of green information systems (GIS) in attenuating the sustainability trade-offs. *International Journal of Production Economics, 255*, 108703.

European Commission. (2020). *INSPIRE Directive implementation report 2020.* Publications Office of the European Union.

Falasco, S., Caputo, P., & Garrone, P. (2024). Can Geographical Indications promote environmental sustainability in food supply chains? Insights from a systematic literature review. *Journal of Cleaner Production*, *444*, 141100.

Fallah, M., Karami, T., & Kojouri, M. K. (2023). The Art Of Human Participation Against Artificial Intelligence With Geospatial Information System In The Tourism Industry. *Geo Journal of Tourism and Geosites*, *49*(3).

Fang, Y., Nie, Y., & Penny, M. (2021). Geographical modeling of COVID-19 transmission using GIS and spatial statistics. *International Journal of Health Geographics*, *20*(1), 1–12. PMID: 33397338

FAO. IFAD, UNICEF, WFP and WHO. (2023). The State of Food Security and Nutrition in the World 2023. Urbanization, agrifood systems transformation and healthy diets across the rural–urban continuum. Rome, FAO. https://doi.org/DOI: 10.4060/cc3017en

Ferguson, R. (2012). Learning analytics: drivers, developments and challenges. International journal of technology enhanced learning, 4(5-6), 304-317.

Folke, C., Carpenter, S. R., Walker, B., Scheffer, M., Chapin, T., & Rockström, J. (2010). Resilience thinking: Integrating resilience, adaptability and transformability. *Ecology and Society*, *15*(4), 20.

Food and Agriculture Organization of the United Nations. Climate Smart Agriculture Sourcebook, Sustainability. *Food Security and Climate Change: Three Intertwined Challenges*. https://www.fao.org/climate-smart-agriculture-sourcebook/concept/module-a1-introducing-csa/chapter-a1-1/en/

Formánek, T., & Sokol, O. (2022). Location effects: Geo-spatial and socio-demographic determinants of sales dynamics in brick-and-mortar retail stores. *Journal of Retailing and Consumer Services*, *66*, 102902.

Geissdoerfer, M., Pieroni, M., Pigosso, D., & Soufani, K. (2020). Circular economy and sustainability: A review. *Journal of Cleaner Production*, *277*, 123741.

Ghosh, P., & Kumpatla, S. P. (2022). GIS applications in agriculture. Geographic information systems and applications in coastal studies, 89-114.

Girard, L. F., & Nijkamp, P. (2019). *Sustainable city and creativity: Promoting creative urban initiatives*. Routledge.

Global Hunger Index. (2024). 2023 Global Hunger Index: The Power of Youth in Shaping Food Systems-Africa Brief.

Goodchild, M. F. (2018). Reimagining the history of GIS. *International Journal of Geographical Information Science*, *32*(12), 1–20.

Goodchild, M. F. (2020). GIScience: The science behind the systems. *Geography Compass*, *14*(3), e12483.

Gopal, S., Staufer-Steinnocher, P., Xu, Y., & Pitts, J. (2022). Semiconductor supply chain: a 360-degree view of supply chain risk and network resilience based on GIS and AI. In *Supply chain resilience: Insights from theory and practice* (pp. 303–313). Springer International Publishing.

Government of India. (2022). *National Geospatial Policy, 2022*. Ministry of Science & Technology.

Gros, B., & García-Peñalvo, F. J. (2016). Future trends in the design strategies and technological affordances of e-learning. In *Learning, design, and technology* (pp. 1–23). Springer.

Gubbi, J., Buyya, R., Marusic, S., & Palaniswami, M. (2013). Internet of Things (IoT): A vision, architectural elements, and future directions. *Future Generation Computer Systems*, *29*(7), 1645–1660.

Guiné, R. de P. F., Pato, M. L. de J., da Costa, C. A., da Costa, D. de V. T. A., da Silva, P. B. C., & Martinho, V. J. P. D. (2021). Food Security and Sustainability: Discussing the Four Pillars to Encompass Other Dimensions. *Foods*, *10*(11), 2732. DOI: 10.3390/foods10112732 PMID: 34829013

Gwet, K. (2002). Computing inter-rater reliability with the SAS system. *Stat Methods Inter-rater Reliability Assess*, *3*, 1–16.

Hair, J. F.Jr, Hult, G. T. M., Ringle, C. M., Sarstedt, M., Danks, N. P., & Ray, S. (2021). *Partial least squares structural equation modeling (PLS-SEM) using R: A workbook*. Springer Nature.

Halder, N., Kumar, M., Deepak, A., Kumar, D., Yasmin, N., Ullah, S., & Al-Ghamdi, S. G. (2025). Spatiotemporal Assessment of Urban Thermal Discomfort in Kolkata, India: Insights from Cloud-Based Remote Sensing. *IEEE Journal of Selected Topics in Applied Earth Observations and Remote Sensing*.

Hammitzsch, M., & Lendholt, M. (2011). Evolving an information logistics database for geospatial early warning systems. *Geomatics, Natural Hazards & Risk*, *2*(2), 95–109.

Ha, N. T., Akbari, M., & Au, B. (2023). Last mile delivery in logistics and supply chain management: A bibliometric analysis and future directions. *Benchmarking*, *30*(4), 1137–1170.

Han, J., Zuo, Y., Law, R., Chen, S., & Zhang, M. (2021). Service quality in tourism public health: Trust, satisfaction, and loyalty. *Frontiers in Psychology*, *12*, 731279. PMID: 34526941

Hargreaves, P. K., & Watmough, G. R. (2021). Satellite Earth observation to support sustainable rural development. *International Journal of Applied Earth Observation and Geoinformation*, *103*, 102466.

Harvey, D. (2019). *Spaces of global capitalism: A theory of uneven geographical development*. Verso.

Herold, M., Goldstein, N. C., & Clarke, K. C. (2003). The spatiotemporal form of urban growth: Measurement, analysis and modeling. *Remote Sensing of Environment*, *86*(3), 286–302.

He, Z., & Turner, P. (2021). A systematic review on technologies and industry 4.0 in the forest supply chain: A framework identifying challenges and opportunities. *Logistics (Basel)*, *5*(4), 88.

Hidayati, D.. (2020). Slum mapping using remote sensing and GIS: Supporting urban resilience. *Cities (London, England)*, *96*, 102–115.

Holling, C. S. (1978). *Adaptive environmental assessment and management*. John Wiley & Sons.

Holmes, W., Bialik, M., & Fadel, C. (2019). *Artificial intelligence in education promises and implications for teaching and learning*. Center for Curriculum Redesign.

Hosen, B., Rahaman, M., Kumar, S., Sagar, L., & Akhtar, M. N. (2023). Leveraging artificial intelligence and big data for advanced spatial analytics and decision support systems in geography. *Malaysian Applied Geography*, *1*(2), 62–67.

Hossain, M., Park, S., & Shahid, S. (2023). Frugal innovation for sustainable rural development. *Technological Forecasting and Social Change*, *193*, 122662.

Hosseini, E., Doaei, Z. S., Jamadi, A., & Yazdani, M. (2024). Examining the effect of resilience on job engagement with the mediation of psychological empowerment and job burnout. *Interdisciplinary Journal of Management Studies*, *17*(2), 475–490.

Hussain, A., Bhatti, M. A., Nawaz, M. A., & Ahmad, T. I. (2019). The influence of geography and its information management system on tourism planning: Mediating role of institutional factor. *PalArch's Journal of Vertebrate Palaeontology*, *16*(1), 1–16.

Hussain, Z., Khan, A., & Ali, A. (2023). The impact of user-generated content, social interactions and virtual economies on metaverse environments. *Journal of Sustainable Economics*, *1*(2), 34–44.

Huyen, N. N. T., Hong, H. D., & Thi, L. H. (2025). Green practices: Building green image and green trust for green revisit intentions in the hospitality industry. *Journal of Open Innovation*, *11*(1), 100481.

IPCC. (2021). [*The Physical Science Basis*. Intergovernmental Panel on Climate Change.]. *Climatic Change*, •••, 2021.

Jat, M. K., Garg, P. K., & Khare, D. (2008). Monitoring and modelling of urban sprawl using remote sensing and GIS techniques. *International Journal of Applied Earth Observation and Geoinformation*, *10*(1), 26–43.

Jeong, H., Karim, R. A., Sieverding, H. L., & Stone, J. J. (2021). An application of GIS-linked biofuel supply chain optimization model for various transportation network scenarios in Northern Great Plains (NGP), USA. *BioEnergy Research*, *14*(2), 612–622.

Joos, G. (2022). Geographic Information Systems in Defense. In *Springer Handbook of Geographic Information* (pp. 685–705). Springer International Publishing.

Kabisch, N.. (2021). Mapping urban nature for sustainable planning: Advances in geospatial methods. *Landscape and Urban Planning*, *214*, 104–136.

Kameswaran, A., Nanthakumar, K., & Ruba, M. (2017). High performance precomputation based self-controlled precharge-free content-addressable memory. *Journal of Electronics and Communication Systems*, 2, 1–9.

Kamini, J., Jayanthi, S. C., & Raghavswamy, V. (2006). Spatio-temporal analysis of land use in urban Mumbai-Using multi-sensor satellite data and GIS techniques. *Photonirvachak (Dehra Dun)*, *34*(4), 385–396.

Kandwal, R., Garg, P. K., & Garg, R. D. (2018). Health GIS and spatial analysis. *Health Informatics Journal*, *24*(2), 228–238.

Karunathilake, E. M. B. M., Le, A. T., Heo, S., Chung, Y. S., & Mansoor, S. (2023). The path to smart farming: Innovations and opportunities in precision agriculture. *Agriculture*, *13*(8), 1593.

KEBEDE. E., 2015. *Land Use/Land Cover Change of Human Factors on the Bene of Lake Abaya* (Doctoral dissertation, Arba Minch University).

Khan, F., Yusoff, R. M., & Kakar, P. (2025). Impact of service quality and customer satisfaction on tourism industry. *Journal of Advanced Research in Social and Behavioural Sciences*, *39*(1), 1–11.

Khan, S. A. R., Godil, D. I., Jabbour, C. J. C., Shujaat, S., Razzaq, A., & Yu, Z. (2025). Green data analytics, blockchain technology for sustainable development, and sustainable supply chain practices: Evidence from small and medium enterprises. *Annals of Operations Research*, *350*(2), 603–627.

Kline, R. B. (2015). *Principles and practice of structural equation modeling*. Guilford publications.

Kochanek, A., Generowicz, A., & Zacłona, T. (2025). The Role of Geographic Information Systems in Environmental Management and the Development of Renewable Energy Sources—A Review Approach. *Energies*, *18*(17), 4740.

Korpinen, O. J., Aalto, M., Kc, R., Tokola, T., & Ranta, T. (2023). Utilisation of spatial data in energy biomass supply chain research—A review. *Energies*, *16*(2), 893.

Krishna, K. S., & Murthy, Y. S. (2015). A study on perception of quality of work life among textile manufacturing workers in east Godavari district. [IJSR]. *International Journal of Scientific Research*, *4*(10), 1487–1491.

Krishna, K. S., & Murthy, Y. S. (2015). A study on perception of quality of work life among textile manufacturing workers in East Godavari district. [IJSR]. *International Journal of Scientific Research*, *4*(10), 1487–1491.

Krishna, K. S., & Murthy, Y. S. (2015). Employee engagement related to employees' performance. *International Journal of Engineering Technology. Management and Applied Sciences*, *3*(8), 153–157.

Krithika, S., Jayachandran, T., Raghapriya, N. R., Ruba, M., Janani, C., & Sathiyavathi, S. (2025, April). An effective denial of service texture features from malware grayscale image. In *Proceedings of the 2025 3rd International Conference on Artificial Intelligence and Machine Learning Applications (AIMLA)* (pp. 1–7). IEEE.

Kumar, L.. (2014). Land degradation mapping using object-based classification in GIS. *Environmental Monitoring and Assessment*, *186*, 8451–8469.

Kumar, S., Singh, R., & Patel, T. (2022). Geospatial technologies for monitoring SDGs. *Sustainable Development*, *30*(1), 123–135.

Kurowska, K., Marks-Bielska, R., Bielski, S., Aleknavičius, A., & Kowalczyk, C. (2020). Geographic information systems and the sustainable development of rural areas. *Land (Basel)*, *10*(1), 6.

Kusdibyo, L. (2022). Tourist loyalty to hot springs destination: the role of tourist motivation, destination image, and tourist satisfaction. *Leisure/Loisir, 46*(3), 381-408.

Lacour, S. O., & Feidt, M. (2022). Entropy analysis in spray cooling for dosing water injection. [). IOP Publishing.]. *IOP Conference Series. Earth and Environmental Science, 960*(1), 012004.

Lata, K. M., Rao, C. S., Prasad, V. K., Badarianth, K. V. S., & Rahgavasamy, V. (2001). Measuring urban sprawl: a case study of Hyderabad. *GIS development, 5*(12), 26-29.

Laurillard, D. (2013). *Teaching as a design science: Building pedagogical patterns for learning and technology*. Routledge.

Lee, M.-T., Liu, J.-M., & Borazon, E. Q. (2020). Evaluating the effect of perceived value of ecosystem services on tourists' behavioral intentions for Aogu Coastal Wetland. *Sustainability, 12*(15), 6214.

Leeonis, A. N., Ahmed, M. F., Mokhtar, M. B., Lim, C. K., & Halder, B. (2024). Challenges of Using a Geographic Information System (GIS) in Managing Flash Floods in Shah Alam, Malaysia. *Sustainability, 16*(17), 7528.

Lehoullier, L. (2023). The Future of Last Mile Delivery: What Consumers are Willing to Sacrifice to Make Sustainable Delivery Decisions.

Lei, Y. (2024). Enhancing environmental management through big data: Spatial analysis of urban ecological governance and big data development. *Frontiers in Environmental Science, 12*, 1358296.

Liping, C., Yujun, S., & Saeed, S. (2018). Monitoring and predicting land use and land cover changes using remote sensing and GIS techniques—A case study of a hilly area, Jiangle, China. *PLoS One, 13*(7), e0200493.

Liu, J., Kuai, X., & Wang, P. (2025). Research on the tourism service quality evaluation of Gongbei Lingnan community under the perspective of SERVQUAL theory. *Frontiers in Sustainable Cities, 6*, 1502234.

Liu, Y., Hou, Y., Guo, J., & Yan, C. (2024). The impact of environment on enhancement of tourism system resilience in China: The moderating role of digital technology. *Technological Forecasting and Social Change, 206*, 123492.

Liu, Y., Huang, D., Wang, M., & Wang, Y. (2020). How do service quality, value, pleasure, and satisfaction create loyalty to smart dockless bike-sharing systems? *Revista Brasileira de Gestão de Negócios, 22*(03), 705–728.

Li, Z., Chen, Y., & Li, B. (2020). GIS-based precision agriculture: A review of methods and applications. *Computers and Electronics in Agriculture, 173*, 105–128.

Logeshwaran, J., Srivastava, D., Kumar, K. S., Rex, M. J., Al-Rasheed, A., Getahun, M., & Soufiene, B. O. (2024). Improving crop production using an agro-deep learning framework in precision agriculture. *BMC Bioinformatics, 25*(1), 341. PMID: 39487390

Longley, P. A., Goodchild, M. F., Maguire, D. J., & Rhind, D. (2011). *Geographic information systems and science* (3rd ed.). Wiley.

Luckin, R. (2018). Machine Learning and Human Intelligence. The future of education for the 21st century. UCL institute of education press.

Lukas, M., & Reis, N. (2022). 1 Peripheral Urbanization: Autoconstruction, Transversal Logics, and Politics in Cities of the Global South. In *Beyond the Megacity: New Dimensions of Peripheral Urbanization in Latin America* (pp. 57-82). University of Toronto Press.

Mahata, D., Shekhar, S., & Ravi, K. (2024). Urban expansion influences on land use and land cover changes in Kolkata metropolitan region: A geo-spatial study. *GeoJournal, 89*(6), 237.

Malakar, K. D., & Roy, S. (2024). Applications of Participatory GIS: A Socio-ecological Approaches and Mapping. In *Mapping Geospatial Citizenship: The Power of Participatory GIS* (pp. 37–48). Springer Nature Switzerland.

Malaker, T., & Meng, Q. (2024). Urban Disparity Analytics Using GIS: A Systematic Review. *Sustainability, 16*(14), 5956.

Malczewski, J. (2006). GIS-based multicriteria decision analysis: A survey. *International Journal of Geographical Information Science, 20*(7), 703–726.

Malczewski, J. (2010). Local spatial decision support systems. *International Journal of Geographical Information Science, 24*(5), 1–25.

Mandal, D., & Ghosh, S. K. (2000). Precision farming–The emerging concept of agriculture for today and tomorrow. *Current Science, 79*(12), 1644–1647.

Masser, I. (2019). *Spatial data infrastructures in context.* CRC Press.

Mathenge, M., Sonneveld, B. G., & Broerse, J. E. (2022). Application of GIS in agriculture in promoting evidence-informed decision making for improving agriculture sustainability: A systematic review. *Sustainability, 14*(16), 9974.

Maurya, A., & Kumar, A. (2024). *The Role of GIS in the Study of Sustainable Development and Environmental Management.* International Journal For Multidisciplinary Research.

Maxner, T., Dalla Chiara, G., & Goodchild, A. (2025). The state of sustainable Urban Last- Mile freight planning in the United States. *Journal of the American Planning Association, 91*(1), 88–101.

Meadows, D. (2011). *Thinking in systems: A primer.* Chelsea Green.

Means, B., Toyama, Y., Murphy, R., Bakia, M., & Jones, K. (2009). Evaluation of evidence-based practices in online learning: A meta-analysis and review of online learning studies.

Meeks, W. L., & Dasgupta, S. (2004). Geospatial information utility: An estimation of the relevance of geospatial information to users. *Decision Support Systems, 38*(1), 47–63.

Merli, R., Preziosi, M., Acampora, A., & Ali, F. (2019). Why should hotels go green? Insights from guests experience in green hotels. *International Journal of Hospitality Management, 81*, 169–179.

Mir, Y. H., Mir, S., Ganie, M. A., Bhat, J. A., Shah, A. M., Mushtaq, M., & Irshad, I. (2025). Overview of land use and land cover change and its impacts on natural resources. In *Ecologically Mediated Development: Promoting Biodiversity Conservation and Food Security* (pp. 101–130). Springer Nature Singapore.

Mittal, G., Bansal, R., Fahlevi, M., & Aziz, A. L. (2025). Mapping the research patterns on fuzzy decision-making trial and evaluation laboratory technique in marketing: A bibliometric analysis. *Social Sciences & Humanities Open, 12*, 101691.

Moise, M.-S., Gil-Saura, I., & Ruiz-Molina, M.-E. (2018). Effects of green practices on guest satisfaction and loyalty. *European Journal of Tourism Research, 20*(20), 92–104.

Morshed, M. M., Hashim, R. A., & Mahmood, R. (2022). Tourist Satisfaction, Destination Image, Loyalty and Revisit Intention to Malaysia's National Parks. *The Asian Journal of Professional & Business Studies, 3*(2).

Murthy, Y. S., Rabby, F., Bansal, R., & Chakravorty, C. (2025). Building Resilient Brands: Harnessing Influencers, Evangelists, and Data-Driven Strategies. In *Data Analytics and Influencer Marketing for Cultivating Brand Evangelism and Affinity* (pp. 71-102). IGI Global Scientific Publishing.

Murthy, Y. S., Sai, K. P., & Murthy, C. R. (2025). Integrating IoB in Smart Education: Innovative Approaches to Behavioral Data Analytics and Pedagogy. In *Internet of Behavior-Based Computational Intelligence for Smart Education Systems* (pp. 237-270). IGI Global.

Murthy, S. Y., Chandra, B. R., & Joshi, A. (2025). Silence is not always golden: Exploring the causes and consequences of employees' silence in the IT sector. In *Digital Transformation and Sustainability of Business* (pp. 283–288). CRC Press.

Murthy, Y. S. N., Chandra, R. B., Chakravorty, C., & Shukla, S. (n.d.). Smartwatches, fitness trackers, and beyond: Evaluating experiential marketing efficacy across different age groups. In *Digital transformation and sustainability of business* (pp. 790–794). CRC Press.

Murthy, Y. S., Bansal, R., Chodisetty, R. C. M., Chakravorty, C., & Sai, K. P. (2025). Transforming minds: AI and machine learning applications in cognitive neuroscience. In *Transforming Neuropsychology and Cognitive Psychology With AI and Machine Learning* (pp. 107–128). IGI Global Scientific Publishing.

Musa, M., Gasmalla, O., Elmahal, A., & Ganawa, E. (2024). *Distributed Geospatial Information Systems Challenges and Opportunities*. Exploring Remote Sensing-Methods and Applications.

Musa, S.. (2015). GIS-based poverty mapping: A global review. *International Journal of Geographical Information Science*, *29*(10), 1–19.

Muyingo, S., Etoori, D., Lotay, P., Malamba, S., Olweny, J., Keesler, K., & Settaala, R. (2022). The procurement and supply chain strengthening project: Improving public health supply chains for better access to HIV medicines, Uganda 2011–2016. *Journal of Pharmaceutical Policy and Practice*, *15*(1), 72. PMID: 36303238

National Food Security Portal, Department of Food and Public Distribution Division, Government of Global. *National Food Security Act (NFSA) 2013*. https://nfsa.gov.in/portal/nfsa-act

Nejat, S., & Fallah Lajimi, H. (2022). Identifying and prioritizing Factors Affecting the Revisit Intention of the Destination Brand in Iran with the Best-Worst Method. *Journal of International Marketing Modeling*, *3*(1), 62–84.

Neuman, M. (2005). The compact city fallacy. *Journal of Planning Education and Research*, *25*(1), 11–26.

Nguyen Huu, T., Nguyen Ngoc, H., Nguyen Dai, L., Nguyen Thi Thu, D., Truc, L. N., & Nguyen Trong, L. (2024). Effect of tourist satisfaction on revisit intention in Can Tho City, Vietnam. *Cogent Business & Management*, *11*(1), 2322779.

Nimmagadda, S. L., Reiners, T., & Wood, L. C. (2019). On modelling big data guided supply chains in knowledge-base geographic information systems. *Procedia Computer Science*, *159*, 1155–1164.

Odion, D., Shoji, K., Evangelista, R., Gajardo, J., Motmans, T., Defraeye, T., & Onwude, D. (2023). A GIS-based interactive map enabling data-driven decision-making in Nigeria's food supply chain. *MethodsX*, *10*, 102047. PMID: 36845370

Pan, C., & Ratasuk, A. (2025). Determinants of Revisit Intention: A Case of Snorkeling Tourism in Hainan Province. *Journal of ASEAN PLUS Studies*, *6*(1), 43–64.

Pandey, P. C., & Pandey, M. (2023). Highlighting the role of agriculture and geospatial technology in food security and sustainable development goals. *Sustainable Development*, *31*(5), 3175–3195.

Parnell, S., & Walawege, R. (2011). Sub-Saharan African urbanisation and global environmental change. *Global Environmental Change*, *21*, S12–S20.

Patel, A. B., & Desai, T. N. (2019). A systematic review and meta-analysis of recent developments in sustainable supply chain management. *International Journal of Logistics Research and Applications*, *22*(4), 349–370.

Pavani, P., Komatireddy, S. R., Yarasuri, V. T., Police, V. R., Tanneru, S. K., & Al-Jawahry, H. M. (2024). Geographic Information Systems Applications in Business Decision- Making. In *E3S Web of Conferences* (Vol. 529, p. 04006). EDP Sciences.

Pham, B. T.. (2022). GIS-based modeling for natural hazard susceptibility using machine learning. *Geomatics, Natural Hazards & Risk*, *13*(1), 345–366.

Plazier, P., Rauws, W., Neef, R., & Buijs, P. (2024). Towards sustainable last-mile logistics? Investigating the role of cooperation, regulation, and innovation in scenarios for 2035. *Research in Transportation Business & Management*, *56*, 101198.

Porter, M. E. (2003). The economic performance of regions. *Regional Studies*, *37*(6–7), 549–578.

Po, S., & Jiang, M. (2023). The effects of green practices on green customer loyalty in the hotel industry: Multi-mediating role of green perceived value and green satisfaction. *International Journal of Research*, *11*(10), 95–106.

Priya, P., & Singh, S. (2022). Factors Influencing the Adoption of Sustainable Agricultural Practices: A Systematic Literature Review and Lessons Learned for Global. *The Forum for Social Economics*, *53*, 1–17. DOI: 10.1080/07360932.2022.2057566

Qu, P. (2024). Design of Supply Chain Logistics Intelligent Management Information System Based on GIS Optimization Model. *Procedia Computer Science*, *243*, 396–405.

Rabby, F., Murthy, Y. S., & Bansal, R. (2025). Brand evangelism in the digital era: The impact of data-driven influencer marketing. *Journal of Open Innovation*, *11*(2), 100552. DOI: 10.1016/j.joitmc.2025.100552

Radianti, J., Majchrzak, T. A., Fromm, J., & Wohlgenannt, I. (2020). A systematic review of immersive virtual reality applications for higher education: Design elements, lessons learned, and research agenda. *Computers & Education*, *147*, 103778.

Rahman, K. T., Bansal, R., & Pruthi, N. (2023). Consumer adoption of technologies: A mindset-oriented approach. In *Contemporary Studies of Risks in Emerging Technology, Part B* (pp. 243–255). Emerald Publishing Limited.

Raihan, A. (2024). A systematic review of Geographic Information Systems (GIS) in agriculture for evidence-based decision making and sustainability. *Global Sustainability Research*, *3*(1), 1–24.

Rajabifard, A., Paudyal, D. R., & Williamson, I. (2020). SDI and good governance. *International Journal of Spatial Data Infrastructures Research*, *15*, 1–15.

Raja, M., Swamy, T. N. V. R. L., Das, S., Bansal, R., Fahlevi, M., & Aziz, A. L. (2025). Does human resource information system influence performance management? *Cogent Business & Management*, *12*(1), 2438862.

Ramaano, A. I. (2025). The essence of geographic information systems (GIS) in sustainable tourism, public leadership and inclusive community participation in remote-African rural societies. *Journal of Responsible Production and Consumption*, *2*(1), 25–47.

Redecker, C., & Punie, Y. (2017). *Digital competence of educators* (Punie, Y., Ed.).

Rejeb, A., Rejeb, K., Abdollahi, A., Zailani, S., Iranmanesh, M., & Ghobakhloo, M. (2021). Digitalization in food supply chains: A bibliometric review and key-route main path analysis. *Sustainability*, *14*(1), 83.

Rezvani, S., Falcão, M. J., Komljenovic, D., & de Almeida, N. M. (2023). A systematic literature review on urban resilience enabled with asset and disaster risk management approaches and GIS-based decision support tools. *Applied Sciences (Basel, Switzerland)*, *13*(4), 2223.

Ritchie, W. J., Kerski, J., Novoa, L. J., & Tokman, M. (2023). Bridging the gap between supply chain management practice and curriculum: A location analytics exercise. *Decision Sciences Journal of Innovative Education*, *21*(2), 83–94.

Rousta, A., & Allaf Jafari, E. (2024). Impact of environmental knowledge, responsibility and concern on sustainable consumption behavior: Does customer attitude matter? *Management of Environmental Quality*, *35*(8), 1858–1877.

Ruba, M., Ponkarthika, M., Parameshwari, V., Girinath, V. P., Soundarya, B., Janani, M., & Santhini, C. (2025, February). Enhancing underwater tunnel safety through smart inspection: An ROV-based system with AI-powered crack detection using Canny edge detection. In Proceedings of the 2025 3rd International Conference on Intelligent Systems, Advanced Computing and Communication (ISACC) (pp. 1067–1072). IEEE. https://doi.org/DOI: 10.1109/ISACC.2025

Ruba, M., Ashwin Prabhu, G., Deka, J., Kavitha, D., Sathiya, S., & Pavithra, S. (2026). Empowering learners and educators: The transformative role of conversational AI in personalized education and student support. In Assayed, S. (Ed.), *Enhancing student support and learning through conversational AI* (pp. 163–190). IGI Global Scientific Publishing., DOI: 10.4018/979-8-3373-3316-8.ch007

Ruba, M., Ashwin Prabhu, G., Robinson Joel, M., Vishnu Priya, K., Dharshani, L., & Dharani, R. (2026). Integrating emerging technologies for organizational agility: An interdisciplinary approach to digital transformation. In Roache, D. (Ed.), *High performance leadership for organizational excellence: Innovation, technology, and resilience* (pp. 63–92). IGI Global Scientific Publishing., DOI: 10.4018/979-8-3373-6207-6.ch003

Ruba, M., Girinath, V. P., Ponkarthika, M., Santhini, C., Sakthi Priya, V., & Hemalatha, R. (2023, December). Design and implementation of smart internet-based real-time control of rover with artificial intelligence. In *Proceedings of the 2023 International Conference on Energy, Materials and Communication Engineering (ICEMCE)* (pp. 1–5). IEEE. https://doi.org/DOI: 10.1109/ICEMCE2023

Sahana, M., Hong, H., & Sajjad, H. (2018). Analyzing urban spatial patterns and trend of urban growth using urban sprawl matrix: A study on Kolkata urban agglomeration, India. *The Science of the Total Environment*, *628*, 1557–1566.

Saini, S., & Bansal, R. (2025). Parasocial relationships and self-disclosure: Strengthening brand image through brand attitude in premium cosmetics. *Journal of Promotion Management*, *31*(2), 337–366.

Sang, L., Zhang, C., Yang, J., Zhu, D., & Yun, W. (2011). Simulation of land use spatial pattern of towns and villages based on CA–Markov model. *Mathematical and Computer Modelling*, *54*(3-4), 938–943.

Santamaria, M., et al. (2020). GIS-based climate change modeling for resilience assessment. *Environmental Research Letters, 15*(9), 094–112.

Santhini, C., Ruba, M., Sakthi Priya, V., & Hemalatha, R. (2025). Intercultural communication and assessment practices in the integration of AI with education. In *Generative AI Approaches to Sustainable Development in Higher Education*. IGI Global.

Satar, M. S., Aggarwal, D., Bansal, R., & Alarifi, G. (2023). Mapping the knowledge structure and unveiling the research trends in social entrepreneurship and inclusive development: A bibliometric analysis. *Sustainability*, *15*(7), 5626.

Saxena, P., Sonwani, S., Gupta, A. K., Shri Rajendra Ratnoo, I. A. S., & Acharya, M. P. (2025). *Natural Infrastructure and Resilient Cities: Role of Green Infrastructure in Urban Planning for Adaptation to Climate Change in India*. National Institute of Disaster Management.

Schönherr, S., & Pikkemaat, B. (2024). Young peoples' environmentally sustainable tourism attitude and responsible behavioral intention. *Tourism Review*, *79*(4), 939–952.

Schultink, G. (2000). Critical environmental indicators: Performance indices and assessment models for sustainable rural development planning. *Ecological Modelling*, *130*(1-3), 47–58.

Scott, A. J. (1995). The geographic foundations of industrial performance. *Competition & Change*, *1*(1), 51–66.

Scott, G., & Rajabifard, A. (2017). Sustainable development and geospatial information: A strategic framework for integrating a global policy agenda into national geospatial capabilities. *Geo-Spatial Information Science*, *20*(2), 59–76.

Selwyn, N. (2019). *Should robots replace teachers?: AI and the future of education*. John Wiley & Sons.

Sen, A. (1983). *Poverty and Famine: An Essay on Entitlement and Deprivation*. Oxford University Press., https://academic.oup.com/book/32827

Seow, A. N., Foroughi, B., & Choong, Y. O. (2024). Tourists' satisfaction, experience, and revisit intention for wellness tourism: E word-of-mouth as the mediator. *SAGE Open*, *14*(3), 21582440241274049.

Shahid, M. K., Khin, A. A., Seong, L. C., Shahbaz, M., & Ahmad, F. (2024). Mapping the relationship of research and development expenditures and economic growth through bibliometric analysis: A theoretical perspective. *Journal of the Knowledge Economy*, *15*(4), 17529–17555.

Shamsuzzoha, A., Ndzibah, E., & Kettunen, K. (2020). Data-driven sustainable supply chain through centralized logistics network: Case study in a Finnish pharmaceutical distributor company. *Current Research in Environmental Sustainability*, *2*, 100013.

Shannon, C. E. (1948). A mathematical theory of communication. *The Bell System Technical Journal*, *27*(3), 379–423.

Sharma, R., & Tyagi, A. (2022). GIS applications in transportation planning. *Journal of Urban Management*, *11*(3), 345–359.

Sieber, R. E., & Johnson, P. A. (2015). Civic open data and crowdsourcing. *Government Information Quarterly*, *32*(4), 409–418.

Siemens, G. E. O. R. G. E. (2005). Connectivism: A learning theory for the digital age. International Journal of Instructional Technology and Distance Learning. Online] retrieved from: http://www. idtl. org/Journal/Jam _05/article01. html

Song, H., & Ma, Y. (2024). Measurement and spatial and temporal characterization of logistics efficiency—Based on a three-stage DEA model. *Heliyon*, *10*(19). PMID: 39397936

Sowmiya Narayanan, K. J., & Manimaran, A. (2024). Recent developments in geographic information systems across different application domains: A review. *Knowledge and Information Systems*, *66*(3), 1523–1547.

Sudhira, H. S., Ramachandra, T. V., & Jagadish, K. S. (2004). Urban sprawl: Metrics, dynamics and modelling using GIS. *International Journal of Applied Earth Observation and Geoinformation*, *5*(1), 29–39.

Sun, O., & Fan, N. (2020). A review on optimization methods for biomass supply chain: Models and algorithms, sustainable issues, and challenges and opportunities. *Process Integration and Optimization for Sustainability*, *4*(3), 203–226.

Sun, Y., & Sun, H. (2021). Green innovation strategy and ambidextrous green innovation: The mediating effects of green supply chain integration. *Sustainability*, *13*(9), 4876.

Supreme Court of Global. *Kapila Hingorani vs State of Bihar on 9 May 2003*. https://globalnkanoon.org/doc/1455798/

Supreme Court of Global. *Kishen Pattnayak and Anr. Vs State of Orissa on 9 January, 1989*. https://globalnkanoon.org/doc/181576930/

Susanty, A., Sari, D. P., Budiawan, W., & Kurniawan, H. (2016). Improving green supply chain management in furniture industry through internet based geographical information system for connecting the producer of wood waste with buyer. *Procedia Computer Science*, *83*, 734–741.

Tang, H., Wang, R., Jin, X., & Zhang, Z. (2022). The effects of motivation, destination image and satisfaction on rural tourism tourists' willingness to revisit. *Sustainability*, *14*(19), 11938.

Thakur-Weigold, B., Buerki, P., Frei, P., & Wagner, S. M. (2022). Mapping the Swiss vaccine supply chain. *Frontiers in Public Health*, *10*, 935400. PMID: 35923971

Tian, R., Li, H., Jia, H., & Gu, R. (2025). The impact of digital infrastructure development on the green transformation of cities: Considering the moderating effect of human capital level. *International Review of Economics & Finance*, •••, 104695.

Tsakiridi, A. (2021). Applications of geographic ınformation systems (GIS) in supply chain management: systematic literature. Int. J Sup. Chain. Mgt Vol, 10(5).

UN-GGIM. (2022). *Future trends in geospatial information management: The five to ten-year vision* (3rd ed.). United Nations Committee of Experts on Global Geospatial Information Management.

United Nations. Human Rights, Office of the High Commissioner (1966). General Assembly Resolution 2200A (XXI). *International Covenant on Economic, Social and Cultural Rights*. https://www.ohchr.org/en/instruments-mechanisms/instruments/international-covenant-economic-social-and-cultural-rights

United Nations. Human Rights, *The Right to Adequate Food, Fact Sheet No. 34*. https://www.ohchr.org/sites/default/files/Documents/Publications/FactSheet34en.pdf

United Nations. Meeting Coverage and Press Releases (13 February 2024). Climate Action Can Help Fight Hunger, Avoid Conflicts, Official Tells Security Council, Urging Greater Investment in Adaptation, Resilience, Clean Energy. https://press.un.org/en/2024/sc15589.doc.htm

United States Environmental Protection Agency US-EPA. (2025). Climate Impacts on Agriculture and Food Supply, Climate Change Impacts. https://www.epa.gov/climateimpacts/climate-change-impacts-agriculture-and-food-supply

Uzairuddin, S., & Jaiswal, M. (2022). Digital monitoring and modeling of construction supply chain management scheme with BIM and GIS: An overview. *Materials Today: Proceedings*, *65*, 1908–1914.

Venkatesh, V., Thong, J. Y. L., & Xu, X. (2012). Consumer acceptance and use of information technology. *Management Information Systems Quarterly*, *36*(1), 157–178.

Viera, A. J., & Garrett, J. M. (2005). Understanding interobserver agreement: The kappa statistic. *Family Medicine*, *37*(5), 360–363.

Vinuesa, R.. (2020). The role of artificial intelligence in achieving the Sustainable Development Goals. *Nature Communications*, *11*, 1–10. PMID: 31932590

Von Bertalanffy, L. (1968). *General system theory: Foundations, development, applications*. George Braziller.

Vujovic, S., Haddad, B., Karaky, H., Sebaibi, N., & Boutouil, M. (2021). Urban heat island: Causes, consequences, and mitigation measures with emphasis on reflective and permeable pavements. *CivilEng*, *2*(2), 459–484.

Wang, Y., & Cao, S. (2011). Carbon sequestration may have negative impacts on ecosystem health.

Wang, B., Hu, D., Hao, D., Li, M., & Wang, Y. (2021). Influence of government information on farmers' participation in rural residential environment governance: Mediating effect analysis based on moderation. *International Journal of Environmental Research and Public Health*, *18*(23), 12607. PMID: 34886344

Wang, L., & Hsu, H. H. (2024, September). The Utilization of Geographic Information System in Logistics Distribution. In *Proceedings of the 2024 2nd International Conference on Internet of Things and Cloud Computing Technology* (pp. 131-135).

Wang, L., Liu, L., Meng, X., Gao, Q., & Fan, M. (2024). The Relationship between Geographical Self-Efficacy and Academic Achievements in Geography: A Moderated Mediating Model. *Sustainability*, *16*(7), 2682.

Wang, X., Su, H., & Liu, X. (2025). The impact of green technological innovation on industrial structural optimization under dual-carbon targets: The role of the moderating effect of carbon emission efficiency. *Sustainability*, *17*(14), 6313.

Wei, H., Wang, Y., Yang, C., & Yu, P. (2024). Empirical Analysis of the Role of Digital Agriculture in Enabling Coordinated Development of Ecosystem Services and Human Well-Being: Evidence from Provincial Panel Data in China. *Sustainability*, *16*(23), 10199.

Weng, Y., Zhang, J., Yang, C., & Ramzan, M. (2024). Intermodal travel planning and decision support integrated with transportation and energy systems. *Heliyon*, *10*(11). PMID: 38828355

Williamson, B., Eynon, R., & Potter, J. (2020). Pandemic politics, pedagogies and practices: Digital technologies and distance education during the coronavirus emergency. *Learning, Media and Technology*, *45*(2), 107–114.

Wilson, A. G. (1971). *Entropy in urban and regional modelling*. Pion.

Wilson, E. H., Hurd, J. D., Civco, D. L., Prisloe, M. P., & Arnold, C. (2003). Development of a geospatial model to quantify, describe and map urban growth. *Remote Sensing of Environment*, *86*(3), 275–285.

Wong, C. B., Law, M., & Wu, W. C. B. (2023). Determinants of customer loyalty: A green marketing perspective. *10th International Conference on Business, Accounting, Finance and Economics*

Wu, F. (2016). Emerging Chinese cities: Implications for global urban studies. *The Professional Geographer*, *68*(2), 338–348.

Wu, J., Gan, W., Chao, H. C., & Yu, P. S. (2024). Geospatial big data: Survey and challenges. *IEEE Journal of Selected Topics in Applied Earth Observations and Remote Sensing*.

Xia, L., Lee, T. J., & Kim, D. K. (2024). Relationships between motivation, service quality, tourist satisfaction, quality of life, and spa and wellness tourism. *International Journal of Tourism Research*, *26*(1), e2624.

Xiaomei, Y., & RongQing, L. Q. Y. (1999). Change Detection Based on Remote Sensing Information Model and its Application on Coastal Line of Yellow River Delta. *Earth Observation Center, NASDA, China*.

Yachai, K., Kongboon, R., Gheewala, S. H., & Sampattagul, S. (2021). Carbon footprint adaptation on green supply chain and logistics of papaya in Yasothon Province using geographic information system. *Journal of Cleaner Production*, *281*, 125214.

Yadav, A., Garg, R. K., & Sachdeva, A. (2024). Artificial intelligence applications for information management in sustainable supply chain management: A systematic review and future research agenda. *International Journal of Information Management Data Insights*, *4*(2), 100292.

Yamijala, S. M. S., Chodisetty, R. C. M., & Chakravorty, C. (2025). Adapting HR practices for a transhumanist future: An empirical study on policy and implementation. In *Enhancing the Modern Workforce Through Transhumanism* (pp. 117–146). IGI Global.

Yang, K. (2024). Integrating geographical information systems into urban-rural planning for equitable and sustainable urban development. *Information Development*, •••, 02666669251339133.

Yashvardhan, V., Pyare, R., & Singh, R. (2022). Technological Advancement in Digital Farming and Global's Present Scenario: A Review. *International Journal of Environmental & Agriculture Research*, *8*(12). Advance online publication. DOI: 10.5281/zenodo.7496469

Ye, H., Brown, M., & Harding, J. (2018). GIS for all: exploring the barriers and opportunities for underexploited GIS applications.

Yeh, A. G. O.. (2017). GIS-based modeling for urban sustainability and smart growth. *Environment & Planning B*, *44*(6), 1012–1035.

Yeh, A. G. O., & Li, X. (2001). Measurement and monitoring of urban sprawl in a rapidly growing region using entropy. *Photogrammetric Engineering and Remote Sensing*.

Yokota, T., & Pagkalinawan, H. (2022). Application of geographic information systems in impact evaluation and geospatial portfolio analysis of transport projects. *Transportation Research Record: Journal of the Transportation Research Board*, *2676*(11), 171–185.

Yousaf, Z. (2021). Go for green: green innovation through green dynamic capabilities: accessing the mediating role of green practices and green value co-creation. *Environmental Science and Pollution Research International*, *28*(39), 54863–54875. PMID: 34018115

Yu, C., Arshad, M. A., Zhao, M., & Yao, W. (2025). How Does GIS Training Affect Turnover Intention of Highway and Bridge Industry Technicians? The Mediating Role of Career Growth and the Moderating Mechanism of Work Anxiety. *Buildings (Basel, Switzerland)*, *15*(15), 2742.

Yusianto, R., & Hardjomidjojo, H. (2020, September). Intelligent spatial decision support system concept in the potato agro-industry supply chain. In *2020 International Conference on Computer Science and Its Application in Agriculture (ICOSICA)* (pp. 1-7). IEEE.

Yu, Y., Yazan, D. M., Bhochhibhoya, S., & Volker, L. (2021). Towards Circular Economy through Industrial Symbiosis in the Dutch construction industry: A case of recycled concrete aggregates. *Journal of Cleaner Production*, *293*, 126083.

Zailani, S., Govindan, K., Iranmanesh, M., Shaharudin, M. R., & Chong, Y. S. (2015). Green innovation adoption in automotive supply chain: The Malaysian case. *Journal of Cleaner Production*, *108*, 1115–1122.

Zaroujtaghi, A., Mansourihanis, O., Tayarani, M., Mansouri, F., Hemmati, M., & Soltani, A. (2025). A Systematic Review of GIS Evolution in Transportation Planning: Towards AI Integration. *Future Transportation*, *5*(3), 97.

Zawacki-Richter, O., Marín, V. I., Bond, M., & Gouverneur, F. (2019). Systematic review of research on artificial intelligence applications in higher education–where are the educators? *International Journal of Educational Technology in Higher Education*, *16*(1), 1–27.

Zhang, Q., Wu, Z., & Guo, G. (2017, April). The cooling effect of water landscape in high-density urban built-up area a case study of the center urban district of guangzhou. In *2017 2nd International Conference on Frontiers of Sensors Technologies (ICFST)* (pp. 394-400). IEEE.

Zhang, S., Wang, J., & Guan, J. (2024). How does the energy consumption structure affect the green economic development? A spatial impact analysis. *Frontiers in Environmental Science*, *12*, 1412612.

Zhang, X., Li, Y., & Chen, J. (2019). GIS-based industrial land suitability assessment. *Sustainability*, *11*(18), 5003.

Zhang, X., Wang, J., Vance, J., Wang, Y., Wu, J., & Hartley, D. (2020). Data analytics for enhancement of forest and biomass supply chain management. *Current Forestry Reports*, *6*(2), 129–142.

Zhang, Y., Shao, W., & Thaichon, P. (2021). Investigating tourist post-travel evaluation and behavioural intention: A cultural intelligence perspective. *Asia Pacific Journal of Marketing and Logistics*, *33*(10), 2037–2053.

Zhao, F., Li, X., & Xu, X. (2023). Integration of AI and GIS for environmental prediction. *ISPRS Journal of Photogrammetry and Remote Sensing*, *197*, 45–60.

Zhao, J., He, T., Xi, X., Li, W. H., & Liu, W. (2025). Geographic proximity and supply chain Resilience: Unravelling their complex dynamics in the digital age. *Technovation*, *148*, 103328.

Zhao, S. X., Guo, N. S., Li, C. L. K., & Smith, C. (2017). Megacities, the world's largest cities unleashed: Major trends and dynamics in contemporary global urban development. *World Development*, *98*, 257–289.

About the Contributors

Arman Khan is an esteemed academician, deeply committed to advancing knowledge and fostering intellectual growth. With over a decade of experience, he has solidified his reputation in the field of academia and research. He holds a Ph.D. from Putra Business School, Malaysia's first AACSB accredited business institution. Currently, Dr. Khan serves as an Assistant Professor of Marketing at Shaheed Benazir Bhutto University, Pakistan. In addition to his teaching role, Dr. Khan has demonstrated a significant presence on both national and international academic stages. His research has been featured in high-impact journals, and he has presented his work at numerous conferences worldwide. Dr. Arman has authored numerous books published by renowned publishers, further showcasing his academic expertise. Through collaboration with fellow scholars, he has refined his research methodologies and expanded his academic contributions. Dr. Khan's editorial roles include serving as Chief Editor of a reputable research journal and being an active member of several editorial boards across national and international journals. His leadership in these roles reflects his dedication to elevating academic discourse and nurturing an environment that encourages innovation and scholarly excellence. Dr. Khan continues to shape the trajectory of academic research, fostering an environment of innovation and excellent.

Zahid Hussain is a Research Assistant the Department of Business Administration. He has a Phd Scholar at Khadim Ali Shah Bukhari Institute of Technology, (KASBIT), Karachi, Pakistan. His research interests include sponsorship, generational marketing, non-profit marketing, marketing communication, consumer behavior and consumer culture. His research publications have appeared in Journal of Retailing and Consumer Services, Journal of Product and Brand Management, Journal of Hospitality and Tourism Research, SEARCH: Journal of Media and Communication Research, Asian Journal of Business Research, Malaysian Journal of Communication, among others. He serves in the editorial board of scholarly journals. Hussain has industry experience in banking and corporate branding before turning to academia.

Abdelrehim Nader Awad is an Assistant Professor of Marketing at the University of Bisha, Saudi Arabia. He holds a PhD in Business Administration (Marketing) from Al-Azhar University and has extensive teaching and research experience in digital marketing, strategic marketing, and customer knowledge management. Dr. Awad has published widely in Scopus and Web of Science indexed journals and currently serves as Vice President of the Innovation and Entrepreneurship Center at the University of Bisha. He is also actively involved in academic supervision, quality assurance, and program accreditation initiatives.

Rohit Bansal is working as an Adjunct Faculty in Rockford College, Sydney, Australia. He obtained Ph.D. in Management from Maharshi Dayanand University, Rohtak. With a rich experience of 17 years, he has achieved growth through robust and proactive academic initiatives. He has authored & edited many books with renowned national & international publishers including Springer Nature, IGI Global, Scrivener-Wiley Publishing, De Gruyter etc. He has published 180 research papers and chapters in national and international journals of repute including Scopus indexed as well as edited books. His research work is published in leading publishers like Springer, MDPI, HIndawi, IGI Global, Cell Press, De Gruyter, Elseveir, Inderscience, Wiley etc. He has also presented papers in 60 conferences and seminars including IIM Indore, IIM Ranchi, IIM Jammu and IIM Kozhikode. He is on Editorial Advisory Board as a member in many national and international peer reviewed journals. He has served as member of technical committee in many international conferences. He has acted as Session Chair and speaker in many international conferences . He has been awarded many times for contribution to academics and research.

Hind Hammouch holds a Ph.D. in Economics and Management from the Faculty of Legal, Economic, and Social Sciences (FSJES) at Sidi Mohamed Ben Abdellah University in Fez, Morocco. She currently serves as a Professor at the Institute for Training in Offshoring Professions in Information Technology, Communication, and Audiovisual (IFMOTICA) in Fez and as a part-time lecturer at FSJES Fez, where she teaches undergraduate and graduate-level courses. Her research interests span corporate social responsibility (CSR), (Environmental, Social, and Governance) ESG, management control systems, organizational and financial performance, as well as the integration of artificial intelligence in sustainable development. Dr. Hammouch has authored and co-authored numerous peer-reviewed articles published in Scopus-indexed journals, as well as scholarly book chapters with IGI Global and other international academic publishers. Her academic contributions also include editorial board memberships and peer-review engagements with several reputable international journals and conferences. With a multidisciplinary background and a strong commitment to applied research, Dr. Hammouch's work contributes to bridging the gap between theory and practice in emerging areas of management, sustainability, and digital innovation.

***

Aishwarya Arya Pursuing Ph.D. in Management, Accomplished her Bachelor's and Master's from Institute of Hotel Management and Catering Technology, Lucknow in Hospitality Administration thereafter another Master's in Travel & Tourism Management. She started her academic career in the year 2019 from IHM Lucknow, previously worked with Babu Banarsi Das University, Lucknow as Assistant Professor. She presented and attended many National and International Conferences, seminars and webinars. Her area of specialization includes Room Division, Hospitality Management & Sales and Marketing. She has patents published in Under the office of the controller General of Patents, Designs and Trademarks Department of Industrial Policy and Promotions, Ministry of Commerce and Industry, Government of India.

Saurabh Chandra is an accomplished legal scholar and currently serves as an Associate Professor & Assistant Dean at the School of Law, Bennett University, India. With a robust academic foundation, he earned his Bachelor of Arts and Bachelor of Laws (B.A. LL.B.) from the prestigious Aligarh Muslim University (A.M.U.). He further specialized in Business Laws, completing his Master of Laws (LL.M.) at the renowned National Law School of India University (NLSIU), Bangalore. Dr. Chandra's academic journey culminated in a Doctorate in Law (Ph.D.) from the esteemed National Law University (NLU) Delhi. In addition to his impressive legal education from some of India's most prestigious institutions, Dr. Chandra also holds a master's degree in management with specialization in Human Resources in Management and another in Journalism and Mass Communication. With almost two decades of teaching experience, he has numerous publications to his credit. His expertise is complemented by a robust portfolio of scholarly publications. By bridging the gap between theoretical insights and real-world applications, Dr. Chandra's work serves as a valuable resource for academics, legal practitioners, and industry professionals alike. Dr. Chandra brings a wealth of experience in Academic Administration, having successfully held various key administrative positions. His diverse qualifications and extensive experience make him a distinguished figure in the academic and legal communities.

Chen Hongmei, Professor, School of Economics and Management, Yanshan University, Qinhuangdao 066004, China. Chen has been on the faculty of Yanshan University since at least 2008 and is listed as a professor in the School of Economics and Management. Her research spans regional economic development, logistics and supply-chain studies (with applications to ports and coal logistics), digital innovation ecosystem resilience, and dynamic path/route optimization for urban transport and sharing-mode systems. Chen publishes in both Chinese and international journals and collaborates frequently with scholars on empirical and applied topics relevant to regional logistics, risk and resilience, and digital/AI-enabled decision-support for management.

Elahe Hosseini has Ph.D. of Business Administration in Faculty of Economics, Management & Accounting at Yazd University in Iran. She is lecture of Yazd university and Shiraz University in IRAN. Her research interest is in Employee voice, Human Resource Management and social and entrepreneurship activities to develop entrepreneurship in developing countries. Also, she has several publications in international journals and participated in several conferences in some countries. The title of her doctoral thesis is about the employee voice of knowledge-based companies in Iran. She serves as an editor, a member of the editorial board in a series of distinguished journals such as the Iranian Journal of Management Studies and International Journal of Management and Enterprise Development. Also, she serves as a reviewer for a series of distinguished international journals such as Sustainability Journal, Journal of Transnational Corporations Review, Journal of Open Sage and etc. ORCID: 0000-0003-3478-5320

Navonita Mallick is a doctoral candidate at KIIT Law School, Bhubaneswar, specializing in Child Rights in India. She is actively engaged in research, publications, and presentations at national and international conferences, addressing critical issues of child protection, digital transformation, and legal reform. She has authored multiple books and scholarly articles published by renowned international publishers, addressing diverse legal themes ranging from criminal law to sustainable governance. Navonita's analytical and communication skills drive her commitment to fostering impactful legal and social change.

Muhammad Asif Qureshi is a distinguished academician, researcher, and industry expert with nearly 15 years of diverse experience in teaching, research, and strategic consulting. He currently serves as Associate Professor and Head of the Research & Business Solution Centre at Bahria University Karachi Campus, where he leads initiatives to promote high-impact research and university-industry collaboration. He is also the CEO of Apex Horizon Consultants, a consultancy committed to bridging the gap between academia and industry through innovative research solutions, international conferences, and student-focused development programs. Dr. Qureshi holds a Ph.D. in Human Resource Management from University Utara Malaysia, an AMBA and AACSB-accredited institution, and completed his Postdoctoral Fellowship at Universiti Malaysia Perlis. Dr. Qureshi has extensive international exposure, having served as a Supervisory Committee member at Putra Business School, Universiti Putra Malaysia, and collaborated on multiple international research projects. His scholarly work is published in prestigious outlets, including the A*-ranked International Journal of Hospitality Management (Impact Factor 11.07), and numerous Scopus Q1, SSCI, SCI, ABDC, ABS, and ESCI indexed journals. He is a recipient of the Emerald Literati Awards for research excellence. He has edited academic books for renowned publishers such as IGI Global and Emerald and serves as a reviewer for various high-ranking international journals. Known for his command over quantitative and qualitative methodologies, as well as AI-driven research, Dr. Qureshi is frequently invited as a keynote speaker and session chair at globally recognized conferences, including events hosted by Oxford University, Harvard, MIT, and academic bodies such as the Academy of Management (USA), EURAM, and BAM. He has successfully supervised over 10 MS and Ph.D. students and is passionate about developing future researchers. His research interests include Human Resource Management, Leadership, Green HRM, Sustainable Innovation, CSR, Fintech, and AI-driven HR Analytics. Dr. Qureshi remains actively engaged in enhancing research quality worldwide through his roles as a reviewer, editor, and mentor.

Shivansu Sachan is Pursuing Ph.D. in Management, a Master's holder from Integral University, Lucknow, Bachelor's from IHM, Lucknow. Academic and professional accomplishments of 6 years with renowned organizations and institutes. My Teaching Career has been very satisfying and have continuously upgraded the skills and knowledge by attending many courses and training programs, seminars, conferences organized by different organizations and also delivered many lectures as a Hospitality sector.he is working as assistant professor at Csjm University, kanpur

Shashwata Sahu, a distinguished research scholar currently pursuing her Ph.D. in Law at KIIT University, Bhubaneswar, embodies an unwavering passion for the field of law. With a profound interest in Constitutional Law, she stands as a self-driven scholar, firmly rooted in the principles of altruism. With an impressive literary portfolio, Shashwata has authored nine impactful books, each delving into crucial aspects of law and society. Her works include "The Growing Significance of Artificial Intelligence," "The role of Indian media in covering crimes: a critical analysis," "The use of Amicus Curiae Briefs at the International Criminal Tribunals," "Assessing the legal landscape of bio-medical waste management in India," "Journey of Law: From Deities to Decrees," "Evolution of Local-Self Government in India", "The Invisible Manual Scavengers of India", "Law and Justice in a Globalizing World: An LL.M. Companion" and "Comparative Public Law: An LLM. Companion." Beyond her authored books, Shashwata's contributions extend to numerous publications in both international and national journals indexed in Scopus and Web of Science.

K Pardha Sai is a result Oriented Management professional with an excellent ability to collaborate with students at all levels. He thrives in challenging situations and follows innovative teaching practices to effectively manage students of various recent trends in finance sector. His primary area of interests are Finance, Commerce and Marketing

Ramesh Chandra Sethi, an Assistant Professor of Political Science and Public Policy at the Kalinga Institute of Industrial Technology (KIIT) University, Bhubaneswar, India, is known for his research that has practical applications. Before joining KIIT University, he worked as an Assistant Professor at the Central University of Jharkhand, Ranchi. He teaches Political Science to Undergraduate students and Systems of Governance and Public Policy to Postgraduate and Ph.D. students at the KIIT University. From 2013 to 2016, he worked at the Centre for Innovations in Public Systems (CIPS), an Autonomous Centre of the Administrative Staff College of India (ASCI), Hyderabad, established and funded by the Government of India. While working at CIPS, Dr. Sethi worked directly with the Department of Health and Family Welfare of various State Governments in India. His research focuses on the theoretical-empirical study of the policy process, systems of governance, public policy, health policy, and rural development. Dr. Sethi has published research papers in reputed national and international journals. Some of his research publications span Law and Public Policy, Public Health, Indigenous People and the Environment, Climate Change, and Sustainable Development Goals.

Neeraj Kumar Singh is a Professor of Business Management in the area of Systems Management, Information Technology, E- Commerce, MIS, Digital Marketing and Advertising with more than 22 Years of experience in Academic, Research and Administration. He is D.Phil from University of Allahabad and has published many research papers and books. Currently, he is on deputation as a Registrar, Harcourt Butler Technical University, Kanpur.

Jahangeer Ahmed Soomro is a PhD Scholar in the School of Economics and Management at Yanshan University, Qinhuangdao 066004, China. His research focuses on sustainable development, environmental economics, technological change, and macroeconomic policy analysis. He has contributed to several peer-reviewed studies exploring themes such as CO_2 emissions, renewable energy, financial development, trade openness, social entrepreneurship, and the socioeconomic implications of AI and automation. His published works include examinations of the drivers of carbon emissions, validations of macroeconomic hypotheses, and analyses of innovation and inequality across emerging economies. Prior to pursuing his doctoral studies, Soomro gained academic and teaching experience in business, finance, and economics, building a strong foundation in empirical research and data-driven policy evaluation. His scholarship continues to advance interdisciplinary understanding at the intersection of economics, sustainability, and technological transformation.

Index

A

C

D

E

F

G

H

I

K

L

P

R

S

T

U